W9-BHB-212

The Impending Dominance of the Electric Car

and Why America Must Take Charge

JAMES BILLMAIER

Advantage®

Published by Advantage, Charleston, South Carolina.
Member of Advantage Media Group.

Printed in the United States of America.

ISBN: 978-1-59932-220-9
LCCN: 2010910047

To my mother, now deceased, who nonetheless
remains the most important voice in my head.

To my father, who "helped" me construct an electric motor for my
sixth grade science project—with a horseshoe magnet, some copper
wire, a 16-penny nail, and some metal scraps—and let me believe
it was all my doing. That was the kind of dad he was, and still is.

To my children, Kris and Britt, and to all future generations.
They deserve better from this generation.

Most of all, to my wife, Michelle, who has been my guidepost
and beacon for 32 years. Because she thinks with her heart,
she is always right about the important things.

FOREWORD

In all the time I've been in the automobile industry—more than half a century now—the electric car has been a shimmering mirage, clearly in view but totally unreachable, perhaps totally unreal. Yes, there have been successful electric cars in the past, at any given moment there have always been a few in service, but the true arrival of the universally accepted electric car has always been promised for "sometime in the future." In ten years? Twenty? Who knew? Mark Twain once remarked, on the subject of fluvial transport, "When it's steamboat time, you steam." Today, it's electric car time.

The advantages of electric cars are multiple and compelling. Instant torque for good initial acceleration, quiet operation, low local pollution, and ease of operation are among their many compelling virtues. Their disadvantage is singular, but even more compelling: they couldn't go very far on a single charge. The reason is simple: the energy density of batteries was too low, the time to recharge them too long. That was true in the nineteenth century, it was true in the twentieth, and it is true today, with one major difference: the zeitgeist is positive, a critical mass of users clearly perceives the many advantages, and the perpetual negative—limited range—is finally correctly seen as irrelevant in the majority of cases.

The revelation that electric cars could be practical despite that range limitation came to me in 1965. On March 13 that year, I bought a new Mustang, trading in the small car that I had, coincidentally, bought on the same date a year before. I have always kept logbooks on my cars, so I had a solid record of what I had done with the departed 38 horsepower Saab 93B. It had never gone faster than 70 mph (because that was its top speed), had only once traveled 100 miles from my home, and any time I had done more than 100 miles in a single day, I could have recharged an electric for at least 8 hours at my destination. So, even with old-technology lead-acid "golf cart" batteries, I could easily have done the 24,000 miles I racked up that year with a primitive electric car.

Note that everyone accepted limited performance from small cars like VW Beetles and my Saab in those days, whereas today 150 horsepower seems to be a minimum requirement. Even so, electric cars have become accepted, and at least three major manufacturers—Audi, Nissan, and Renault—are publicly committed to volume production of pure electrics. I'm certain the venturous buyers of plug-in electrics will find them perfectly satisfactory for more than 90 percent of their driving needs, but the critical requirement is that we all realize that fact and behave accordingly. Right now many people are perfectly willing to drive three-ton SUVs all the time because they "might need the space when we go on vacation." Rationally, everyone should rent suitable alternate vehicles for the few times range or load-carrying capability is vital and requirements exceed those of a private EV (electric vehicle).

The principal advantage for everyone in adopting electric cars is virtually eliminating urban air pollution. Few remember that gasoline cars, now seen as villains, were once hailed for freeing city dwellers from the disease-producing organic pollution of horses. The civil advantages of electrics were recognized early. Thomas Edison created a pretty good nickel-iron battery a hundred years ago, and his good friend Henry Ford was on record as planning to make inexpensive and practical electric cars as early as 1915. It didn't happen because the Edison battery was not quite good enough to allow electrics to prevail over the clear advantages of gasoline engines then.

Today's battery technology is vastly better, and advancing rapidly. A wide variety of efficient, lightweight electric motors is available to car makers. An infrastructure to support recharging is being extended in many urban areas, not just in the U.S. but in both established and emerging economies around the world. What this means for the United States is that there is an opportunity to once again lead the world by providing safe, practical, and inexpensive vehicles. Ninety years ago more than half of all the cars in the world were made in the U.S., with consequent prosperity, good jobs, and a thriving domestic economy. If the U.S. embraces the potential of electric vehicles, we can achieve world leadership again. And we can totally eliminate the importation of petroleum products. We have more than enough for all purposes if we don't burn it up unnecessarily in our outmoded twentieth-century vehicles. *JOLT!* points the way, with details on every aspect of the electric vehicle revolution.

—Robert Cumberford
Automotive Design Editor, *Automobile* Magazine

CONTENTS

INTRODUCTION

A couple of years ago some colleagues and I, all computer-technology veterans from Silicon Valley's earliest days, were discussing the state of world affairs over a couple glasses of wine. (Okay, maybe a few.) We agreed that while we had helped build technologies and industries that have provided a transformational foundation for the planet and its people, we sure had left our kids with a lot of daunting challenges.

Wouldn't it be nice, we asked, if we could actually do something to eliminate at least one of these problems?

But where would we start? What could we do?

After much discussion, we agreed energy independence was the biggest threat to America's national security. We agreed that the United States could no longer afford to rely on overseas oil to power its economy. Dependence on foreign petroleum is simply too expensive—from a strategic perspective, an economic perspective, and a national security perspective. The U.S. sends hundreds of billions of dollars overseas each year to support our insatiable need for oil, much of which comes from countries that despise us. We also spend hundreds of billions of taxpayer dollars annually to protect that oil—and not just for us, but for the rest of the world as well.

But we couldn't do anything about it, could we?

I am pretty sure it was somewhere around the second bottle of wine that we concluded that yes we could. And yes we would.

How? We would do what we have always done. We would find a better technology.

Lucky for us, that technology appeared to already exist—in the form of the electric vehicle (EV). We decided to check it out, to see if the technology met our expectations. If it did, we would work to transform the nation's transportation system from one based on oil to one based on electricity, which is made here in America. Replacing the cars we drive would go a long way toward helping the U.S. achieve energy independence and increased national security—not to mention boosting our economy and helping the planet along the way.

So after 30 years in the computer systems and software technology industries, I changed tack. I had always been interested in energy issues. Even before leaving my post as a math teacher and entering the computer industry in 1980, I had followed early-generation electric vehicle advances, hydrogen fuel cell research, and battery and other energy storage developments.

But after that night I dove in headfirst. For the past two years I have spent much of my time studying all aspects of the next-generation electric vehicle. And I have come to understand that we're looking at a technology and economic revolution bigger than the computer and Internet revolutions combined, with the EV the "killer app" that will launch this next wave.

The internal combustion engine is dying. Its death throes may take 20 years, but make no mistake: the end is coming. And that's an excellent thing, since as you'll read in *JOLT!*, EVs represent a better, faster, and cheaper mode of transportation. Ending our nation's reliance on foreign oil and helping the planet along is great. But the real reason EVs will come to dominate the personal transportation market—cars, SUVs, vans, and pick-up trucks—over the next couple of decades is that they make financial sense to the consumer. Bottom line: they are cheaper to operate and maintain than gas-powered vehicles. (And as you'll learn, they're an absolute blast to drive.) Just as consumers ultimately powered the computer and Internet revolutions, consumers will propel the EV revolution as well. Americans will adopt EVs in overwhelming numbers—in the process driving yet another paradigm shift of massive proportions.

Electric vehicles also offer a phenomenal business opportunity. While the Internet represents an annual $1 trillion market worldwide, legendary Silicon Valley venture capitalist John Doerr has projected that EVs and the associated energy market will be six times bigger, accounting for $6 trillion a year worldwide. Speaking before a Senate committee in 2009, Doerr told members that energy technology "is the mother of all markets, perhaps the biggest economic opportunity of the twenty-first century."

The great unknown, however, is whether or not the U.S. will be prepared to profit from the EV revolution. The coming "electriconomy"—an economy based on an electrified personal transportation system—will result in both massive upheaval and massive opportunity. China, in particular, has acknowledged the inevitability and the potential of the EV revolution and is in fast-forward mode to implement the new technology. But the electriconomy is as essential to America's national security as is energy independence, and Chinese ownership of the EV realm would leave the U.S. in a dangerous position. Possessing the technologies that power our economy is crucial to America's strength and well-being.

There is no longer any question of whether or not we will adopt an electric-based transportation system. We will. And the transition will come much more quickly than most "experts" predict. All major automakers have some type of plug-in vehicle coming out in the very near future, with the first cars due out at the end of 2010. The U.S. can't afford to be left behind. But we're going to need to move fast to become the undisputed market leader.

The good news is that we're halfway there, at least in terms of ability. The U.S. has a well-established history of economic leadership and is renowned for its innovation. It also has a resourceful and skilled workforce able to capitalize on every aspect of the coming electriconomy, from conception and development to manufacture and delivery. In short, the U.S. workforce is a veritable Dream Team.

And the electric vehicle is a Dream Car. EVs are good for us individually. They're good for us as a nation. And they're good for the planet. Hang on, America! The EV is going to take us on an amazing ride.

—James Billmaier

PART I

Driving the EV Highway

"Everything in life is somewhere else, and you get there in a car."

—E.B. White, American author

"We choose to go to the moon in this decade...because that goal will serve to organize and measure the best of our energies and skills, because that challenge is one that we are willing to accept, one we are unwilling to postpone, and one which we intend to win."

—John F. Kennedy, 1962

CHAPTER 1

Sputnik Redux

On October 4, 1957, the Soviet Union launched a silver orb the size of a beach ball into space and changed the course of history. The first artificial Earth-orbiting satellite, the 184-pound Sputnik sported long, whiskery antennas and—for 22 days until its transmitter battery died—emitted a lively-sounding beep that was picked up by awestruck scientists and ham radio operators more than 500 miles below. Circling the globe every 96 minutes, the shiny sphere was a declaration of Soviet technological superiority, with the United States as the intended target.

Americans focused their binoculars and telescopes on the flash of light streaking by at 18,000 miles an hour and knew the world was changing before their eyes. What weapons did the satellite carry? What messages were being relayed to its masters half a world away? What did it mean for the future?

For those not yet born or too young to remember, it's hard to imagine the sense of fear that gripped the nation with the launch of Sputnik. The world was then in the early stages of the Cold War, with the U.S. and the U.S.S.R. staring each other down across a line of fear and hostility—a line backed on both sides by huge arsenals of nuclear weapons. *Popular Mechanics* magazine ran a cover story called "How to Build a Backyard Bomb Shelter." Schoolchildren ran duck-and-cover drills.

With an intensity bordering on paranoia, Americans responded to Sputnik by focusing on the space race. Such focus, however, required a national mood shift. Isolationism had gripped the U.S. in the years following World War II. We had won, after all; we felt we deserved to be left alone to build new cars, new houses, and new lives in communities that later came to be known as the suburbs. Life was good. We had Dwight D. Eisenhower in the White House, Howdy Doody and Ed Sullivan on TV, easy credit from the banks, and cheap gas at the pump.

Sputnik blew that carefree isolationism to bits.

Americans today are as disengaged and complacent about the threats to our national well-being as we were in 1957. We're more self-absorbed than ever, happily inhabiting a world in which reality television and Facebook posts command more attention than the Iraq and Afghanistan wars. Unless you're related to a soldier, the conflicts in the Middle East don't seem to matter much.

But that sense of complacency is once again being challenged. Roaring at us from the western horizon is a technology revolution so big it could dwarf nearly everything we accomplished in the years after Sputnik. This time around, though, the threat isn't just limited to the issue of military dominance. Today the United States faces a new threat to its national security. This time the threat is one of economic dominance, and it comes in the form of the electric car.

The electric car?

Yes, the electric car. Though not as dramatic or obvious a threat as Sputnik, with its potential to aim missiles at America's cities and heartland, the threat is no less real, and the danger far more insidious. The EV industry is certain to become a technological juggernaut. Any society that masters it will enjoy an economic bonanza and enormous world power.

For our own security, that society needs to be us.

Make no mistake: the electric vehicle is coming. We will all be driving EVs in 20 years—and many of us will be driving them this year—for some simple but very compelling reasons: They are much cheaper to run than gas-powered cars. They are much cheaper to maintain. And because they rely on electricity, an energy source produced here in the U.S., they are cleaner. They're also a blast to drive, offering more power, torque, and acceleration than a comparable gas-powered car delivers.

From a national standpoint, EVs represent yet another outstanding benefit: the opportunity to break our addiction to oil. Two thirds of the oil we use is imported, much of it from countries hostile to the United States. That makes our dependence on foreign oil as much a threat to our national security as falling to second—or even third—place on the economic rung. And in the global economy of the twenty-first century, energy independence and economic power are increasingly and intrinsically connected.

The question is, will the technologies that sustain the explosive new EV economy—the coming electriconomy—be made in the USA, or will they be made in China?

Unlike our flat-footed stance at the outset of the space race, the U.S. already has the EV technologies needed to compete. But technology know-how is not enough. We also need national passion and focus to ensure those technologies are designed, developed, and manufactured in the United States of America. A homegrown electric vehicle industry is essential to restoring and safeguarding our independence, security, and global strength.

We've done it before, and we can do it again. Americans are pioneers, inventors, and trailblazers. It's in our national DNA. When our backs are up against the wall, we do what we need to do.

Three months and 37 million miles after launch, Sputnik fell from orbit and disintegrated in Earth's atmosphere. But those three months had far-reaching consequences. Sputnik kicked the space race into high gear. The Soviets' triumph was a massive wake-up call, shocking the U.S. out of its smug perception of itself as the technologically superior nation. (We had cars with tailfins, after all; theirs looked like the boxes our cars might come in.) Caught off guard and driven by fear of military attacks from the sky, a stunned and frightened America dove into action.

In an all-hands-on-deck burst of energy, Americans mobilized around a common goal: to regain the technological lead. For the next 10 years the nation's attention turned to scientific research. Mathematics and science departments saw a surge in enrollment. The Department of Defense went into overdrive on missile development. Congress initiated the National Aeronautics and Space Administration (NASA), lighting the fuse on 50 years of innovation and a new economy of products and industries. Entire, never-before-imagined industries grew out of the space program, including silicon chips, the personal computer, the Internet, kidney dialysis, the CAT scanner, water purification technologies, and anything and everything digital. (And although NASA didn't invent the space-age, orange-flavored drink Tang, it did make it famous.) In the end, despite the fears it engendered, Sputnik captured the country's imagination, inspired careers, propelled industries, and transformed Americans' views of science and education.

But what, exactly, is it we need to do today?

We need to get to work. We need to build our own EV manufacturing base—everything from the cars we drive to the charging infrastructure and smart-grid systems that support them. Many countries, including Germany, France, and Japan, have acknowledged the inevitability and the potential of EVs, and are in overdrive to use and profit from the new technology. But it is China in particular that sees electric vehicles and the associated alternative energy industries as a way of transforming itself into a nation of innovation, while at the same time maintaining its lead in global manufacturing.

In and of itself, China's goal of an electrified transportation sector poses no threat to the United States. But dependence on China or any other nation for EV technology—from the vehicles themselves to the components that power and support them—simply replaces one vulnerability with another, still leaving us open to massive economic disruption.

That risk to our national security is as dangerous as our current reliance on foreign oil.

And if we're all sick of hearing about the decline of America and the rise of China, why not move to assure our energy security now? Why not work to shape and benefit from the new economy? Why not step up and do whatever it takes to be a leader on the world economic stage for decades to come?

Why would we ever surrender this battle—let alone surrender it in advance?

Many American companies are already invested in various aspects of China's new technologies, so the odds are good that there will be some degree of cooperation and cross-pollination between the U.S. and China in their pursuit of the EV market. Yes, it is a global market, and interdependencies will continue for mutual benefit. But cooperation does not mean the end of competition and the need for strategic self-reliance. The fallout surrounding the recent China-based hacker attacks on Google's trade secrets and email accounts is proof that the technology competition between the U.S. and China is anything but benign. As China emerges as an economic and military superpower, it has become crystal clear that the U.S. cannot afford to leave the core infrastructure of EV technology in the hands of any country other than our own.

There's something else to consider, something equally important to our nation's well-being. How's the economy in your part of America? Much of our great country is still reeling from what's being called the Great Recession, and economists warn us that even if the recession is over, at least technically, the era of high joblessness is just beginning. Unemployment and underemployment—which includes people who've given up looking for work or who are working fewer hours than they'd like—hit 17.4 percent in late 2009, the highest level since the 1930. Even more stunning, a recent survey revealed that nearly 45 percent of American families had suffered a job loss, a pay cut, or reduced hours in the past year.

But here's the good news, America: EVs offer a way out. According to the Electrification Coalition's 2009 policy guide on adopting plug-ins, "The electrification of transportation represents the next great global manufacturing industry, with the potential to bolster economic growth and create American manufacturing jobs."

It's perfect. Ultimately, the EV revolution will be consumer-driven, and the U.S. leads the world in consumerism. We love to buy new things; we are trend-savvy and quick to embrace technology. Pushing forward on EVs offers a way to "consume" the country into a world leadership position in a critically important industry—a way to create an expanding economy along with a more secure nation. As with Sputnik, however, it's show time. We can't afford to stand by and miss out on the alternative energy revolution, which Doerr (who funded EBay, Netscape, and Google) has described as the biggest economic opportunity of the twenty-first century.

The race for who will drive and benefit from the EV revolution is just getting started, and the outcome is far from certain. While other nations, including China, already offer hybrid or all-electric vehicles, the first reasonably priced, mass-produced American-made plug-in car, the GM-Volt, will be introduced in late 2010, with the already-sold-out Nissan LEAF, an American-manufactured, full-battery electric vehicle, due out around the same time. (The U.S.-based Tesla Motors has had a limited-production, $100,000-plus electric sports car on the market for the past two years.) Nonetheless, when it comes to EVs, China faces production and technology challenges of its own. In the game of EV baseball, we are tied in the early innings.

But America needs a comprehensive energy strategy to win. As Doerr has argued, the biggest obstacle to widespread investment in the clean tech market is the fact that the government has yet to indicate strong interest in alternative energy, much less tie its success to the nation's future prosperity. Although the U.S is renowned for its technological innovation—while China thus far is not—innovation in the alternative energy arena is continually stymied by a lack of clear national direction, which has resulted in tentative private capital investment.

The Pew Environment Group reports that in addition to China, the U.S. has fallen behind 10 other countries, including Canada, Mexico, Turkey, Brazil, Italy, and the U.K., in alternative energy investments as a percentage of the national economy. "The United States' competitive position is at risk in the emerging clean energy economy," said Phyllis Cuttino, director of the Pew Environment Group's Global Warming Campaign, in a statement about the study. "Our nation has a critical choice to make: pass the federal policies necessary to position us as the world leader in the large and growing global clean energy market or continue to watch as China and other countries race ahead." This isn't a call for Big Government; it's simply a call for national energy and innovation policies that can unleash the unmatched power and leadership of American industry.

If we're going to win the EV race, we need to get moving. Now. Leadership in the EV industry will require a collective push, just like the one after Sputnik. It will take everything from the implementation of government incentives to renewed support for math and the sciences, both of which have been greatly eroded since the days of our post-Sputnik successes. Our response to the Soviet threat produced a better educated and more innovative and enterprising nation, as countless students filled science and engineering classrooms throughout the late 1950s and 1960s, eager to help build a stronger, more secure America. Even more, the cutting-edge technologies that grew out of Sputnik formed the core of what would become a booming economy, leading to the decades of America's greatest growth and prosperity.

The moral then—as now—was that when threatened and pushed to the wall, Americans respond with initiative and excellence. We don't try, we do. Pragmatism—the philosophy that says if it works, it's true—is the American way.

When President George W. Bush grabbed a bullhorn from a New York City rescue worker and climbed atop the rubble of the World Trade Center three days after the terrorist attacks of 9/11, he gave the speech of his presidency. Except for one thing. There was no mention of the role that imported oil played in the attacks on America. There was no mention of the need to break our nation's dependence on that oil. And beyond a few references to the importance of breaking our "addiction" to oil in later—much later—speeches, the president essentially told us to resume our lives as before. In retrospect, we can see that was a huge mistake—a lost opportunity. Remember how wounded we all were then? Americans would have followed our commander in chief just about anywhere, particularly if his calls to action promised to move the nation toward safety and security. Imagine where we'd be today if on September 12, 2001, the president, backed by 300 million Americans, had committed the nation to energy independence by 2010.

So far President Obama has done little better; in fact, he's been anything but bold. Like his predecessor, Obama has occasionally spoken of the need to end America's reliance on imported oil, but he has thus far committed less than 0.1 percent of the United States budget in the form of grants to accelerate EV adoption. And he has yet to initiate a comprehensive plan for their deployment, repeatedly passing up golden opportunities to unite the nation around the nonpartisan goal of oil independence.

"President Obama seems intent on squandering his environmental 9/11 with a Bush-level failure of imagination," Thomas Friedman wrote in his *New York Times* column nearly a month after the devastating April 20, 2010, BP oil rig explosion and resulting spill in the Gulf of Mexico. "He is offering no big strategy to end our oil addiction."

But it is not too late. Obama can still give the speech that Bush did not. He can take a page from JFK's famous Moon speeches in 1961 and 1962, when the president issued passionate calls for the U.S. to go head-to-head with the Soviet Union's space program and beat our rivals to the Moon. He can call upon American ingenuity, technical expertise, and can-do spirit—good old-fashioned American grit—to make the electrification of our personal transportation sector, and the energy independence that comes with it, a national imperative.

On July 20, 1969, 12 years after the Soviets launched Sputnik and just 8 years after President Kennedy challenged the nation to send a man to the Moon, Neil Armstrong stepped off Apollo 11's ladder and onto the soft lunar surface, taking "one giant leap for mankind."

Looking back, maybe you think the boot prints on the Moon were most important. Or perhaps you believe the many technologies spawned by our race to the Moon made our lives better. Whatever your beliefs about the space program, however, take note: the impact of the EV revolution can and will be much greater. And yes, it would be great if our nation's leaders stepped up to the plate and worked together to accelerate an American-based EV revolution. But whether they do or not, it's time for us to get off the couch and hit the road.

"We can evade reality, but we cannot evade the consequences of evading reality."

—Ayn Rand, Russian-American philosopher and novelist

CHAPTER 2

NATIONAL INSECURITY

Just how important is oil to the United States? The answer, in black and white, on the U.S. Department of Energy website:

"Oil is the lifeblood of America's economy."

Oil supplies over 40 percent of the country's energy needs, 30 percent of U.S. industry, and almost 100 percent of the fuel needed to power our cars and trucks. Such reliance on oil to support our economy and our lifestyles makes the United States the world's primary oil consumer as we burn through some 20 million of the 85 million barrels produced worldwide each day.

But that huge thirst for oil creates a dangerous situation in terms of security. Because the U.S. holds just 2 percent of the world's oil reserves, two thirds of the oil we consume is imported—almost 14 million barrels each and every day. That's nearly enough oil to fill 360 Olympic-size swimming pools—a line of pools that if placed end-to-end would stretch for 10 miles. And although the U.S. is home to just 4 percent of the world's population, those 14 million barrels of imported oil add up to more than the total daily oil consumption of any other nation.

AMERICA, WE HAVE A PROBLEM

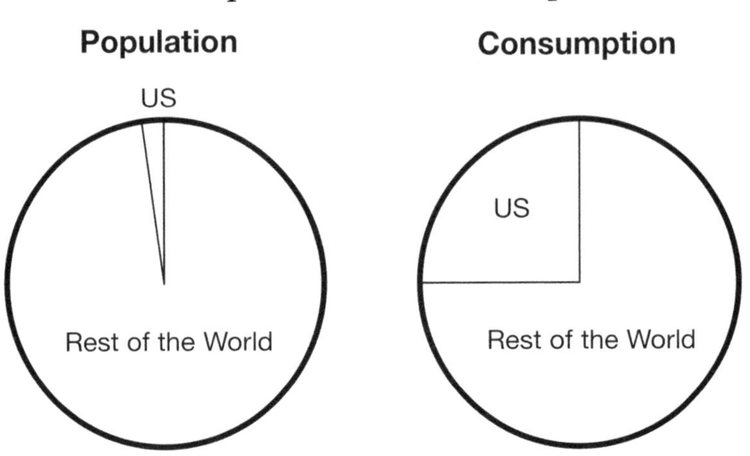

With 300 million inhabitants, the U.S. represents just 4 percent of the world's population yet consumes nearly 25 percent of its oil supply.

But dependence on imported oil leaves us reliant on a small number of countries, many of them hostile to U.S. interests or located in politically unstable regions of the world. No wonder a 2005 survey by the nonpartisan Civil Society Institute showed that two thirds of Americans feel it patriotic to buy a fuel-efficient vehicle. Statistics from the U.S. Energy Information Administration reveal that while 20 percent of our imported crude oil and petroleum in 2008 came from Canada, a slightly larger percentage originated from Persian Gulf countries, with Saudi Arabia accounting for as much as 14 percent. Much of the remainder came from Venezuela, Nigeria, and Mexico, with a total of 6 million barrels of crude oil imported daily from OPEC nations.

TIME TO CHANGE THE GAME

Worldwide Supply of Oil and Oil Reserves by Country

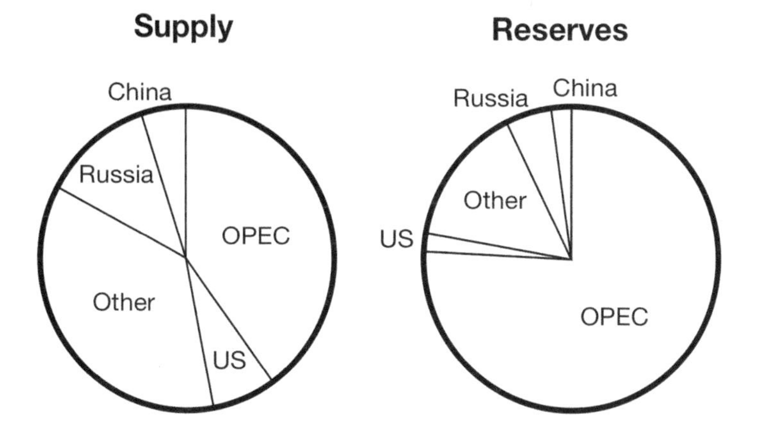

Since the U.S. controls less than 2 percent of all known oil reserves, Americans are dependent on foreign oil, most of which comes from nations that are hostile to American interests or located in politically unstable parts of the world.

"Keeping America competitive requires affordable energy," President George W. Bush declared in his 2006 State of the Union address. "And here we have a serious problem: America is addicted to oil, which is often imported from unstable parts of the world."

Given the overwhelming importance of petroleum to the American economy, any disruption to the flow of oil would have devastating effects. Transportation alone accounts for over 70 percent of total U.S. oil consumption. Cars and light trucks consume two thirds of that amount, which equates to over 20 gallons of gas a week, 52 weeks a year, for every household in America.

To ensure unimpeded delivery of oil to the world market, the U.S. has for many years committed its military to the protection of global oil production and trade, spending as much as $83 billion annually to patrol shipping lanes and defend vulnerable infrastructure from terrorist attack in unstable areas. The Department of Energy notes that every day over half the world's oil supplies must pass through one of six coastal chokepoints, including the crucial Strait of Hormuz, which sits between Iran and Qatar. An unsuccessful attack on this or any other transit point would send oil prices soaring; a successful attack and even brief closure would cause a worldwide economic crisis.

President Obama echoed George W. Bush in his own speech on energy legislation in 2009. "For more than three decades we've talked about our dependence on foreign oil, and for more three decades we've seen that dependence grow," he said. "There's no disagreement over whether our dependence on foreign oil is endangering our security. We know it is."

Further proof that this is a bipartisan issue comes from T. Boone Pickens' PickensPlan, the renowned energy tycoon's website, which envisions an end to our reliance on foreign oil: "If we are depending on foreign sources for nearly two thirds of our oil," it states, "we are in a precarious position in an unpredictable world."

Yet U.S. dependence on overseas oil shows little signs of abating, and has in fact climbed to record levels in recent decades. Once the worldwide leader in oil production and an oil-exporting nation, the U.S. now imports as much as 70 percent of the oil it uses today, up from just 28 percent in 1973, according to the Foreign Trade Division of the U.S. Census Bureau. Even more troubling, Department of Energy records show that in the five years after the attacks of 9/11, imports from the Middle East jumped 10 percent over the previous five years.

Oil dependence has a long reach, stretching beyond military policy and into the realm of international relations. Efforts to ensure the steady flow of oil often tie our hands diplomatically, affecting foreign policy decisions across the globe, from Russia to Iran to Venezuela. And dependence on oil poses a serious threat to American national security in other ways as well. With some $475 billion annually sent overseas for oil supplies, U.S. dollars end up financing petro-dictators, funding nuclear proliferation, and bankrolling terrorists.

In a 2007 interview with the online magazine *The Futurist*, James Woolsey, a former CIA director under President Bill Clinton who also served as energy and climate advisor to John McCain during his 2008 presidential campaign, pointed to the fact that the U.S. was on track to borrow some $320 billion dollars that year to import oil—nearly a billion dollars a day. A share of that money, Woolsey said, would ultimately end up in the hands of followers of Wahhabism, an austere form of Islam that is the dominant faith in Saudi Arabia and elsewhere, including parts of Pakistan. "The result is that the war on terror is the only war the United States has fought, with the obvious exception of the Civil War, in which we pay for both sides," he said. "This is not a good plan."

"The Senate is now considering increasing government subsidies for corn growers to produce more ethanol. If we produce enough ethanol we can postpone our next invasion of a Middle Eastern country for two to three years."

—Jay Leno, comedian

CHAPTER 3

Gas Costs What?

News flash, America: Every time you fill your gas tank, you're paying about $5.28 a gallon. The sign may read $2.78 (or $3 and up, if you live in California) for self-serve gas, but that posted price doesn't even begin to reflect the reality behind the cost of imported oil.

So where does that missing $2.50 go? Actually, it's not missing; it's just hidden, since the true cost of gas is poorly represented by the price of oil on the world market or in what American consumers see at gas stations. Rather than forking over that extra $2.50 at the pump—at $5.28, a full tank of gas would set the average driver back more than $105—we pay it indirectly, as tax dollars, which go to the U.S. military to police global oil production and supply lines in unstable regions of the world. (Remember the $83 billion that funds U.S. military patrols in unstable regions of the world that are crucial to the flow of oil, including the Strait of Hormuz?) We also pay it in federal subsidies to oil companies, which have received between $15 billion and $35 billion each year from the federal government in the name of research and development, exploration, and other industry costs.

THE TRUE COST OF GAS

Price Per Gallon—Includes Oil Supply Defense and Subsidies to Oil Companies

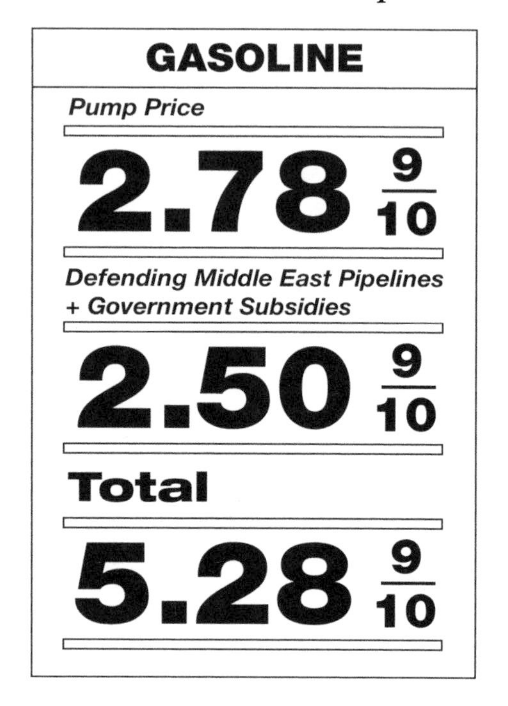

Consumers would pay approximately $5.28 at the pump if the military costs of protecting oil supply lines in unstable regions of the world and government subsidies to oil companies were included in the price. These hidden gas costs are instead paid in tax dollars.

If $5.28 a gallon seems like a lot, think again. It's actually pretty cheap when you consider how much drivers pay elsewhere. Most Europeans, for example, shell out around $7 per gallon. Although the price of raw gas is pretty much the same, European governments tax gasoline at a much higher rate, mostly to fund alternative energy investments and programs to reduce traffic and pollution. While U.S. local, state, and federal taxes make up about 15 percent of the national average retail price of regular gasoline, taxes in France and the U.K. account for around 70 percent. In fact, there has been no increase in the federal gas tax since 1993, even though the cost of gas since then has nearly tripled.

Gasoline in America is a phenomenal deal, one that is too good to last. According to the German Agency for Technical Cooperation, the U.S. is ranked 101st in terms of national gas prices—meaning we pay less for gasoline than do citizens of a hundred other nations. When you consider the affordability of fuel by factoring wages worldwide, statistics show that only Saudi Arabia and Venezuela offer cheaper gas.

ONLY SAUDI ARABIA AND VENEZUELA HAVE MORE AFFORDABLE GAS

Gas Price Comparison by Country

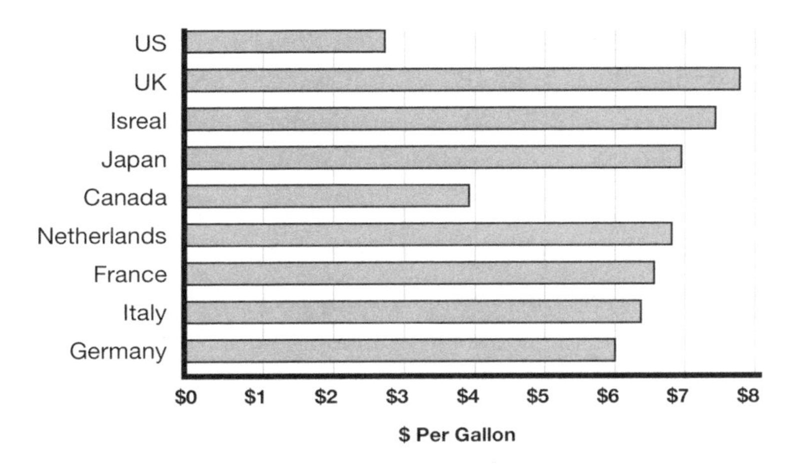

The United States enjoys some of the lowest gas prices in the world. More than 100 nations have higher gas prices than the U.S., including several European countries with gasoline prices exceeding $7 per gallon.

Given that the U.S. imports up to 70 percent of the oil it uses, it doesn't take a degree in economics to see that not only does dependence on foreign oil affect national security, it also puts our nation's economic health at risk. "We're witnessing the greatest transfer of wealth in human history," testified T. Boon Pickens in a recent appearance before Congress. He added that the U.S. trade deficit for January 2010 was $37.3 billion, 74 percent of which was sent overseas to import oil. The PickensPlan website reports that Americans paid approximately $475 billion for foreign oil in 2008. And when oil prices reached $140 a barrel that year, the U.S. was spending some $700 billion for foreign oil. Those hundreds of billions of dollars to feed our oil habit year after year add up pretty dramatically: the Institute for the Analysis of Global Security estimates that almost $1.2 trillion has been transferred to oil-producing nations over the past 30 years, with much of that oil money ending up in the hands of those eager to do us harm.

Thanks to oil prices averaging almost $100 a barrel, OPEC walked away with a record $971 billion in net oil export revenues in 2008. Saudi Arabia raked in the largest share of those earnings, banking almost $290 billion, or 30 percent of all OPEC earnings. The U.S. Energy Information Administration (EIA), an autonomous intergovernmental organization responsible for assessing world supplies for energy importing states, estimates that OPEC members could pocket $767 billion in oil revenues in 2010, and as much as $823 billion in 2011. Worse, unless major steps are taken to curb the world's seemingly insatiable appetite for fossil fuels, EIA anticipates oil prices climbing to almost $200 a barrel in 2030, with more than half of those supplies expected to come from OPEC.

And what about jobs? The Department of Energy reports that every $1 billion of the U.S. trade deficit as a result of buying imported oil equates to the loss of 27,000 jobs. Since oil imports account for nearly one third of the trade deficit, that makes our reliance on foreign oil a big contributor to unemployment, as well as to slowed economic growth, stifled innovation, and an increasingly limited ability to compete on the world stage.

WHY ARE MILLIONS OF AMERICANS UNEMPLOYED?

American Jobs Lost Due to Oil Imports

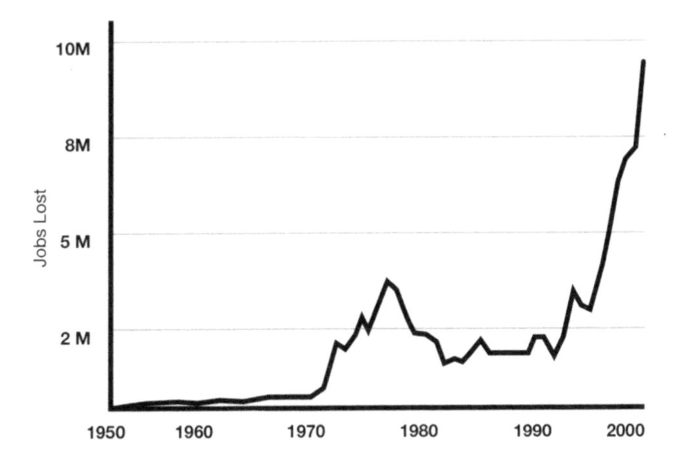

Oil imports account for nearly one third of the trade deficit, making our reliance on foreign oil a major contributor to unemployment. Government figures show that every $1 billion of the U.S. trade deficit as a result of buying imported oil equates to the loss of 27,000 jobs. If the U.S. were able to recapture those jobs, the unemployment rate would drop to below 2 percent.

Countries dependent on a steady supply of oil are always at the mercy of events beyond their control. A disruption in just one part of the oil-producing world—a hurricane in the Gulf of Mexico, guerrilla attacks in Nigeria, political uncertainty in Venezuela, a pipeline accident anywhere—can lead to devastating price fluctuations on the global market. Even a fractional dip in production of less than 10 percent causes wild spikes in spot pricing. And no matter how fast retail gas prices jump, the move back down tends to be nothing short of glacial.

There's more, though. Any interruption to the flow of oil has an almost immediate effect on our pocketbooks. Sudden price hikes hit businesses and households hard, since more cash goes to fuel rather than to growth or consumption.

Probably the biggest effect of an upswing in the price of oil is on consumer spending, which makes up 70 percent of the American economy. More money spent on gas means less money for other things, from a Hawaiian vacation to the purchase of a new refrigerator to dinner at a favorite Italian restaurant. It's certainly no accident that a spike in the price of oil preceded each of the recessions to hit over the past 35 years. And while recessions stem from any number of causes and reach into all sectors of the economy, oil price fluctuations tend to hit fuel-hungry industries the hardest, including airlines, shipping companies, and the auto industry.

All told, researchers at the Oak Ridge National Laboratories estimate that oil dependence has cost the U.S. economy $5.5 trillion between 1970 and 2008, with 2008 alone seeing a loss of $600 billion. That's a lot of money that hasn't been available for things that really matter, things like education, infrastructure needs, law enforcement, technology, and business investment.

In addition, the U.S. has so far spent over $1 trillion on the wars in Iraq and Afghanistan. That works out to over $3,000 per second. Imagine if that same amount of money were instead used to offset the purchase of 50 million electric cars for U.S. citizens, removing 20 percent of gas-powered cars from the road. If we did that, we'd eliminate the need to buy oil from Saudi Arabia and the rest of OPEC.

Now *that's* a weapon of mass construction.

$1 TRILLION AND COUNTING

Total Costs for the Iraq and Afghanistan Wars Annually

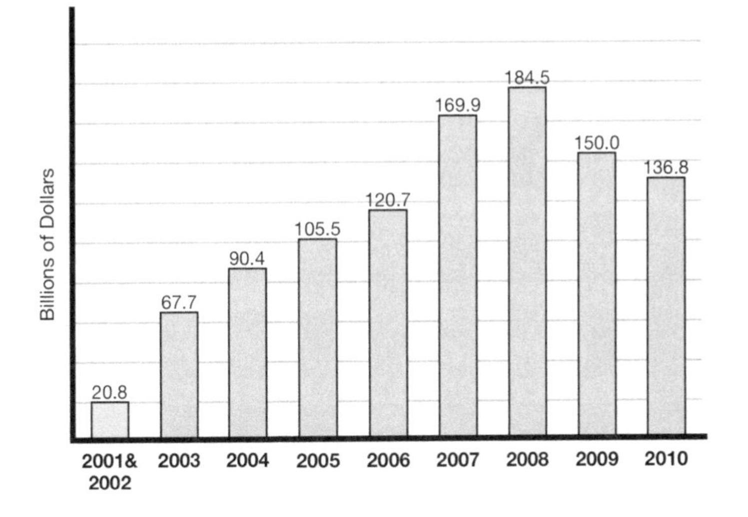

The U.S. has spent more than $1 trillion in taxpayer money to fund the wars in Iraq and Afghanistan. For the same $1 trillion, the U.S. could purchase 35 million electric cars, fund EV infrastructure nationwide, eliminate the need for Middle Eastern oil, and bring our troops home.

Finally, while we obviously can't attach a price to the thousands of lives lost, damaged, or merely interrupted by the war in Iraq—4,412 U.S. soldiers killed and 31,874 wounded as of July 20, 2010, not to mention a $1 trillion outlay thus far—it's worth taking a moment to consider how many fewer conflicts the world might see if we were not forced to safeguard vulnerable global oil supplies.

PAYING THE ULTIMATE PRICE

The Washington Post Faces of War

Some of the U.S. service members who have been killed in Operation Iraqi Freedom and Operation Enduring Freedom.

Visit http://projects.washingtonpost.com/fallen/ *for more information.*

"We've embarked on the beginning of the last days of the age of oil. Embrace the future and recognize the growing demand for a wide range of fuels or ignore reality and slowly–but surely–be left behind."

—Mike Bowlin, chairman and CEO of ARCO (now BP), 1999

CHAPTER 4

Running On Empty

The origins of the fossil fuels—oil, coal, and natural gas—that sustain modern life began more than 300 million years ago. As the remains of plants and animals sank to the bottom of the many oceans, lakes, and swamps of an earlier era, the decomposing matter mixed with mud and microbes. Sand and silt settled atop the muck, and sediment eventually hardened into rock. The passage of time saw the accumulation of even more rock, its massive weight pushing the substance deep into the earth's crust. That pressure worked in tandem with bacteria, heat, and time to transform the material into the carbon-rich fuels that today provide 85 percent of the world's energy.

Cheap and abundant petroleum has stoked phenomenal economic growth worldwide for the past 150 years, and oil plays a crucial role in nearly every aspect of modern industry, from agriculture and food production to the development of pharmaceuticals and anything plastic. The problem, however, is that oil is a finite and nonrenewable resource, one that will eventually run out. When that will happen is debatable but it's clear that in the coming decades oil will become harder to find and more expensive to produce.

In the mid-1950s, M. King Hubbert, an American geologist working for Shell Oil, graphed worldwide oil discoveries and realized they tended to follow a bell-shaped curve. He extrapolated that annual oil production, the amount of oil pulled from the ground and refined in any given year, would follow the same curve. As new reserves were tapped, production would rise to a peak, before falling as all sources were gradually depleted.

Hubbert made his findings public, but Shell had little reason to worry about any fallout—no one paid attention. In fact, just about anyone within the American oil industry who happened upon Hubbert's theory over the next 15 years quickly dismissed it. Never had so much oil been extracted from beneath American soil, each year seeing more production than the last, and the assumption was that the bounty would continue unabated.

But it didn't. Hubbert predicted that annual production in the lower 48 states would peak between 1965 and 1970; in actuality, production peaked around 1970, and has been on a steady decline ever since. (Oil production in Alaska didn't kick into gear until the late 1970s, so the top there came later, in 1988.)

According to IEA, a combination of geologic, technical, and political realities have seen the worldwide supply of oil dropping nearly 7 percent in recent years. The organization released a study in 2009 showing that of more than 800 oil fields comprising three quarters of all global reserves, most of the world's biggest fields have already peaked.

Predictions specific to the Middle East, home to over half the world's oil reserves (and two thirds of all proven reserves), are difficult to make, since state-owned oil companies are cagey when it comes to the issue of sliding production levels. The most optimistic scenarios envision peak oil not hitting the region for another 20 years; the most pessimistic suggest it passed that point around 2005. In a 2005 energy report, for example, ExxonMobil predicted world oil production would peak around 2010.

The only way we'll know for sure is in hindsight. Either way, in addition to the fact that no big oil fields have been discovered in the Middle East in decades, there's little question that some of the longest-running fields are already on the decline. In 2006, Saudi Aramco acknowledged that its mature fields were seeing an annual decline of 8 percent per year, leading experts to believe that Saudi Arabia's Ghawar oil field—the world's largest, which has provided approximately half the country's oil production for the last 50 years—had peaked. Kuwait's Burgan field, the world's second largest, appears to have topped out in 2005. Closer to home, Mexico's enormous Cantarell field, once the mainstay of that nation's oil production, has lost almost half its production capacity in recent years.

While peak oil doesn't mean the end of oil, it does mean the end of cheap oil. Easily accessible petroleum is always extracted first. What remains—often far out at sea or deeply embedded within rock—becomes increasingly difficult and expensive to produce, requiring greater amounts of energy to bring it to the surface and beyond. Back in 1999, energy expert Dick Cheney, then head of Halliburton, one of the world's largest oil field service companies, addressed the ramifications of peak oil. Speaking before the London Institute of Petroleum, he noted that oil production "is obviously a self-depleting activity." Referring to the additional 50 billion barrels of oil he predicted the world would need by 2010, he asked, "So where is the oil going to come from?"

Good question. It took millions of years for oil deposits to form, and it will take a long time to replace what we have so rapidly consumed over the past 200 years. We've been consuming oil since the beginning of the Industrial Revolution, with most of the burn rate occurring in the past few decades. Further compounding the problem is that global demand for petroleum will climb as much as 50 percent between now and 2030—from 85 million barrels to over 130 million barrels a day—all to accommodate the surge in demand coming from developing countries hungry to feed their burgeoning economies and lifestyles. China and India are expected to absorb almost two thirds of that increased production, with one third of that amount expected to fuel the Chinese transport sector alone. In 2004, for example, China's oil demand shot up almost 17 percent.

Meeting the world's predicted energy needs in the coming decades means the earth will ultimately need to spit out more than 40 million additional barrels per day. But where will that oil come from? More specifically, can the pace of new drilling and extraction technologies keep up with the decline in mega-oil field production, given that world oil discoveries have been falling since 1964? Meeting the world's projected oil needs for 2030 will require an increase of 150 percent of all known reserves—half again as much as has already been discovered since the earliest days of modern oil exploration. Such an enormous jump in production would require new oil discoveries comparable to six Saudi Arabias.

NEEDED: SIX MORE SAUDI ARABIAS

Projected Increase in Worldwide Demand for Oil Over 20 Years

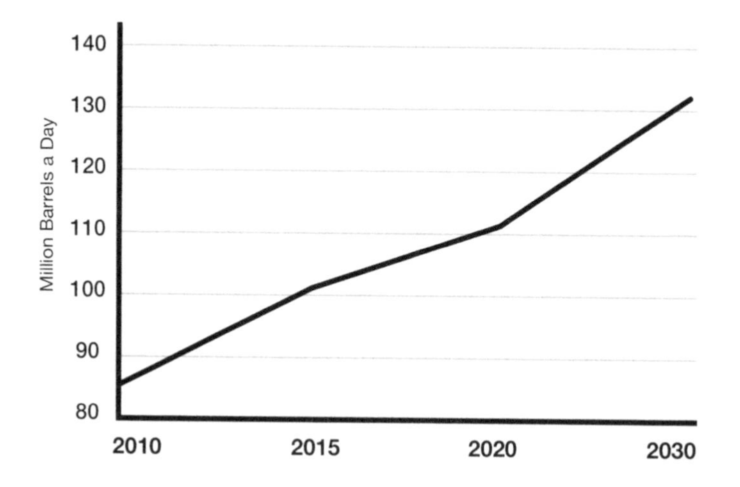

Worldwide demand for oil is expected to climb to over 130 million barrels of oil per day over the next 20 years, an increase of over 50 million barrels a day. Meeting the world's projected oil needs for 2030 will require an increase of 150 percent of all known reserves. That's the equivalent of all the oil fields in Saudi Arabia—times six.

That's a lot of oil, especially when you consider that no such discoveries are on the horizon. And for those convinced that the only energy policy that makes sense for America is "drill, baby, drill," it's important to understand that while the U.S. consumes one quarter of the world's petroleum—some 3 gallons per person daily—it holds just 2 percent of the world's reserves.

While the call for drilling on U.S. soil represents an understandable desire for American energy independence, the unfortunate truth is that domestic drilling can't achieve its intended goal. Not even close. In fact, it's a fool's mission. Estimates of possible offshore and Arctic National Wildlife Refuge oil discoveries range from 0.6 billion to 13 billion barrels, according to the U.S. Department of Energy. Even if successfully tapped, these sites would provide as little as a one-month supply of oil, with a best-case scenario of a two-year supply of oil.

Technological advances in recent years have allowed companies to extract previously unreachable oil from such places as North Dakota's Bakken Shale deposits, which the United States Geological Survey estimates could harbor as much as 4.3 billion barrels, making it the biggest oil discovery in the U.S. in 40 years. But these oil formations produced just 80 million barrels of oil in 2009—enough to supply the U.S. for just four days. It's clear that supplies like these can do little to satisfy the nation's annual habit of more than 7 billion barrels.

These are not energy strategies. They are desperate tactics.

Even if a vast pool of oil in a "friendly" region were out there waiting to be discovered, it could take up to a decade of exploration, study, and construction to get a new rig up and running, depending on where the oil is and how challenging it is to find, test, and develop. The U.S. Department of Energy estimates that offshore drilling in the outer continental shelf, for example, couldn't even begin before 2017, and would take until 2030 to reach peak production (at which point it would provide only 200,000 barrels a day, approximately 1 percent of the country's current demand).

But there's another reason why increased drilling isn't the answer. On April 20, 2010, as this book was being written, the BP oil rig Deepwater Horizon exploded 41 miles off the gulf coast, killing 11 and injuring 17. As *JOLT!* went to press, three months after the explosion, tens of millions of gallons of oil have flowed into the ocean, with uncertainty remaining about the efficacy of the cap now in place and the options available to permanently plug the leak.

The economic and environmental effects of this catastrophe haven't even begun to be estimated, but there is no question that the Gulf of Mexico and the millions of residents along its coast have been greatly impacted. A fishing ban on the most heavily polluted region—the gulf waters provide more than a third of the country's fish and shellfish—has affected hundreds of thousands of commercial and recreational fishermen, and tourism operators along the length of the gulf coast are reporting cancellations due to the spill.

In 1970, when American oil production was at its peak, the U.S. pumped just half of what it consumes today. With demand for energy increasing, the easy oil gone, and our most productive fields in decline, there's no way we can drill our way out of this problem—a conclusion that's been reached by experts across the energy policy divide. Oil veterans like T. Boone Pickens and George W. Bush have both stated publicly that there simply isn't enough petroleum buried beneath American soil, onshore or off, to come close to providing what it takes to run our economy. Which is why after some 60 years in the oil and gas business, Pickens—who has been outspoken on the inevitability of peak oil—has shifted his sights to the development of alternative forms of energy.

That's called looking at the big picture, something we all need to start doing in a hurry.

AMERICAN OIL FIELDS PEAKED IN 1970

U.S. Domestic Oil Production and Oil Imports

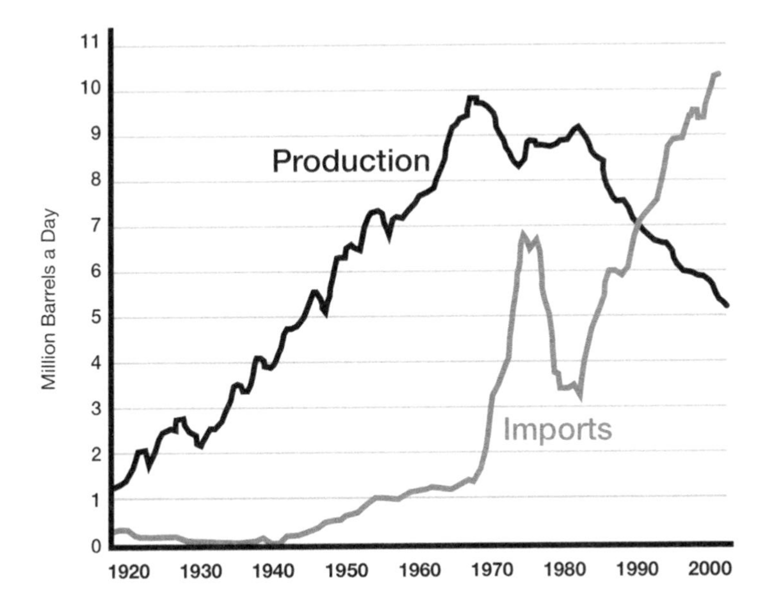

Although the U.S. was once the No. 1 oil producer in the world, many of our most productive oil fields have run dry. What remains represents just 2 percent of the world's known reserves. Even when U.S. oil production was at its peak in 1970, American oil fields produced just half of the 21 million barrels of oil our nation now consumes daily.

"The facts are there that we have created, man has, a self-inflicted wound through global warming."

—ARNOLD SCHWARZENEGGER,
GOVERNOR OF CALIFORNIA

CHAPTER 5

Emissions Overload

There are many smart, well-intentioned people who do not believe that human-produced carbon emissions are a significant contributor to climate change. Some are thought leaders—experts in their fields—and several were interviewed for this book (see Part II).

As an experiment, place yourself on the "Al Gore Chart," which appears on the following page.

(See chart on page 54)

WHERE ARE YOU ON THE "AL GORE CHART"?

Global Warming Matrix

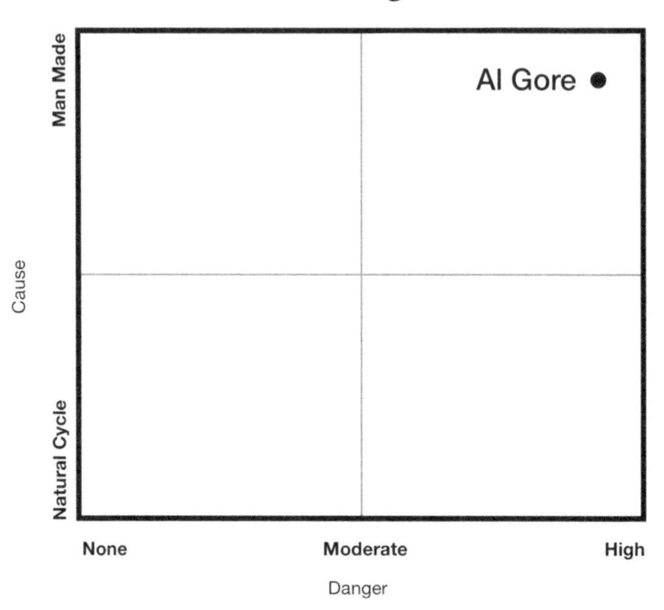

American beliefs vary widely regarding the existence and dangers of global warming, and what—if anything—is the cause.

The above litmus test grew out of a discussion at a Nissan LEAF demonstration event with a Boeing engineer who didn't believe in global warming, and who claimed that the science behind climate change is corrupt. He argued that mankind's ability to affect the atmosphere is *de-minimis*—essentially immeasurable. A heated discussion ensued, with no hope of resolution, until the following question was posed to him: "Why the hell are you here?" The engineer then rattled off a half dozen reasons why electric cars are the future for America, from energy independence to superior handling and performance. It quickly became clear that EVs are a transcendent issue, one with many motivators. Electric vehicles are something Glenn Beck, Rush Limbaugh, Jon Stewart, and Keith Olbermann all can agree on—though perhaps for very different reasons.

If you don't find yourself with Al Gore in the upper-right-hand quadrant of the global warming matrix, or you're just sick of hearing about climate change in general, feel free to skip this chapter. There are plenty of other compelling and less controversial reasons to embrace EVs.

Okay, you're still here.

One of the best things about electric vehicles is that the cars themselves produce zero carbon emissions.

While some may question the validity of the science behind climate change, the evidence is nonetheless solid. Extensive studies conducted by reputable institutions and supported by an overwhelming majority of the scientific community clearly show that global warming is accelerating as a result of man-made emissions. The World Bank reports that humans contribute some 80 percent of the CO2 added to the atmosphere annually, with emissions more than tripling in the last decade compared to the 1990s.

And since cars and light trucks currently account for almost two thirds of U.S. transportation sector emissions, according to the U.S. Environmental Protection Agency, American driving habits are the most significant contributor to the problem. U.S. Census Bureau figures show that although Americans make up just 4 percent of the world's population, they are responsible for nearly half of the world's automotive emissions and a quarter of all greenhouse gas emissions. U.S. carbon dioxide emissions average more than 20 tons per person annually, according to EIA, a number that is six times larger than the global average. We may not want to see the evidence, which points to our mobile lifestyles as having a direct hand in the problem—the United Nations calculates that the average commuter in California contributes more CO2 every year than a single person in Cambodia does over a lifetime—but it's there, and the repercussions are potentially catastrophic if we don't act quickly to turn it around. (And it's not simply a matter of developed versus developing nations. Per capita, for example, the U.S. each year produces twice as many emissions as Germany.)

AMERICA RANKS FIRST IN EMMISSIONS

Per Capita CO2 Emissions

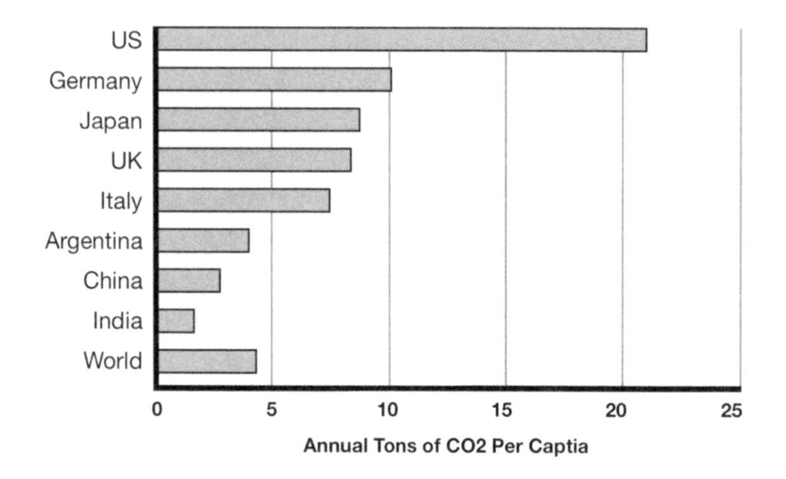

Each American produces more than double the emissions compared to citizens of other industrialized nations.

World CO2 emissions are projected to jump almost 40 percent by 2030, according to EIA. In fact, while the United States still contributes the most greenhouse gases per capita, scientists believe that China has now passed us to become the world's overall leader in carbon emissions. Today, just 10 countries contribute almost 70 percent of all fossil-fuel-related carbon emissions, with the U.S. and China alone accounting for over a third, according to the Earth Policy Institute.

Concern about climate change spans the political spectrum, from tree huggers and carbon counters to military experts and intelligence authorities. It also extends to one of the largest and most established oil companies in the world. Shell Oil—where M. King Hubbert was working when he conceived his concept of Peak Oil more than 50 years ago—prominently posts the following statement, excerpted here, on its website: "For Shell, the debate on climate change is over...Managing CO2 emissions from coal, oil, and natural gas is crucial in tackling climate change."

Although the climate has always cycled in and out of warming and cooling periods—due in part to changes in the earth's orbit and, as a result, the amount of sun hitting different parts of the planet—the issue today grows out of how quickly man-made pollutants are causing global temperatures to rise. Carbon dioxide is an unavoidable byproduct of the fossil fuels burned to run our cars, cook our food, and heat our homes. So-called greenhouse gases act as a blanket, trapping heat in the air surrounding Earth and, as a result of warming global temperatures, disturbing the planet's delicate balance. Signs of strain include:

- Melting icecaps and rising seas
- Long-term drought in some areas and flooding in others
- Unpredictable rainfall, posing a threat to food production in many regions
- More-extreme weather-related events, including tropical storms, heat waves, and wildfires
- An increasing number of plant and animal species headed toward extinction

In terms of global surface temperature, a NASA analysis shows 2009 tied for second place—along with five other years spanning 1988 to 2007—as the warmest year on the books since record keeping began in

1880. Its average temperature was just a fraction of a degree cooler than 2005, which has the dubious distinction of holding the top spot. The year also closed out the warmest decade ever recorded.

WARMEST DECADE ON RECORD

Average Annual Global Increase in Temperature—1980-2010

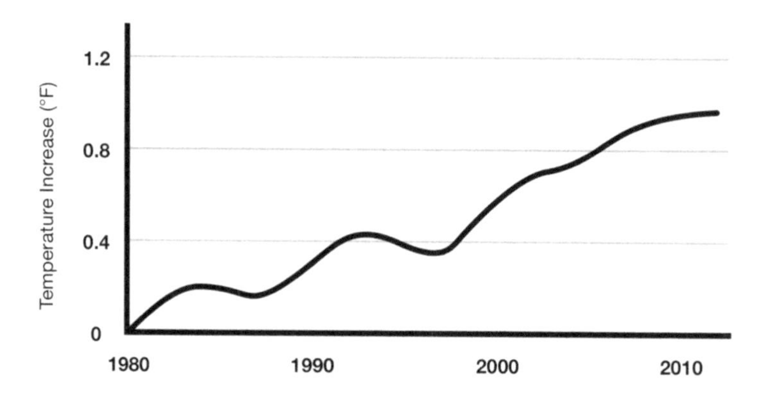

While weather fluctuates from year to year, average global temperatures have been rising for more than a century. Readings from 7,200 thermometers located around the world (both on land and at sea) show that the most recent decade was the warmest on record. Such increases are consistent with projections made by scientists linking a warming planet to increases in CO2 gases.

But don't confuse weather with climate change. Although many Americans saw cooler than average temperatures and unusually harsh weather in December 2009 and January 2010, given that the contiguous 48 states make up just 1.5 percent of the world's land, U.S. temperature fluctuations actually have little impact on the global measurement. Despite heavy rain and thick snowfall throughout parts of the country, the first month of 2010 still managed to achieve the status of second warmest January worldwide on record. Scientists also point out that the extreme weather we saw this past winter actually indicates climate change in action: higher global temperatures result in increased evaporation from the oceans and other bodies of water. As more moisture is pushed into the atmosphere, it eventually falls as additional rain or snow when a storm blows through.

A quick note on the difference between climate and weather: While climate affects weather, they are not the same thing. The difference is the measure of time. Weather is whatever is going on outside at any given moment—a short-term snapshot of current atmospheric conditions. Climate, on the other hand, is the average weather in a particular location over a long period of time.

In any case, it is clear that weather patterns are becoming increasingly erratic, and not just in terms of longer, hotter summers and shorter, warmer winters. Just as Washington, D.C., was digging itself out of an unseasonably severe snowstorm in February 2010, for example, the organizers of the Vancouver Winter Olympics were furiously hauling in snow by the truckload to compensate for the rain that fell in advance the games, which were held at the height of the Canadian winter.

While the temperature increases may seem small at first glance—since the late nineteenth century, average global temperatures have increased by about 1.5 degrees, with much of the change coming in recent decades—just a few degrees difference in any one area can have an outsize effect. The results of a 20-year study, initiated by the first President Bush and released in mid-2009 by the U.S. Global Change Research Program, showed that temperatures in parts of the country have climbed an average of 7 degrees, affecting everything from human health to agricultural output.

Average temperatures in the Arctic have climbed at twice the rate of the worldwide average, according the multinational Arctic Climate Impact Report. As a result, the ice is rapidly receding, and if conditions continue unchecked, the region may well see an ice-free summer around 2040. Glaciers around the world are experiencing an alarming melt rate, with serious long-term repercussions given the crucial role that runoff and seasonal snowmelt play in supplying water to over a billion people. And with the world's population projected to swell from 6.8 billion to 9.2 billion by 2050, that number will certainly increase.

Increasing CO2 levels are also acidifying the ocean's coral reefs, which are second only to tropical rainforests in plant and animal diversity. Reef-building coral are sensitive to even the smallest chemical and temperature fluctuations; 1988, for example, a warmer-than-average year, saw bleach rates of 70 percent in some reefs, and scientists expect many more stress-related die-offs in the coming decades if water temperatures continue to climb.

Of course, when it comes to the ocean, warming temperatures are not the only issue. The explosion aboard the Deepwater Horizon oil rig in the Gulf of Mexico highlights yet another downside to our dependence on oil. The resulting oil spill, the worst in U.S. history, has fouled beaches and environmentally fragile wetlands along the Louisiana, Mississippi, Alabama, and Florida coastlines. Prior to the placement of the cap, an amount equivalent to the 1989 Exxon Valdez disaster—when 250,000 barrels of oil surged into Alaska's formerly pristine Prince William Sound—is estimated to have poured into the gulf every four days.

Scientists also report that giant plumes of oil have spread deep within the gulf, including one 10 miles long, 3 miles wide, and 300 feet thick, further threatening the gulf marine environment and underwater food chain. Chemical dispersants to break up the oil have been used in unprecedented amounts, despite the fact that they themselves are highly toxic, and the long-term effect of such heavy use is unknown.

In addition to their devastating effects on the planet, climate change and other man-made disasters such as the gulf spill also pose a threat to national security. In its 2010 Quadrennial Defense Review, the U.S. Department of Defense states that climate change is already taking place, and warns of the challenges facing the U.S. military in terms of its readiness and ability to respond to the effects of more frequent natural disasters, rising sea levels, long-term drought, and pandemics, as well as to food, water, and resource security. The Pentagon report outlines the strain that would be imposed on the armed forces in the face of volatile, Somalia-like conditions in unstable regions of the world that are important, even essential, to U.S. interests. Intelligence agencies paint a grim picture of the chaos that could stem from climate-related stress: Failed states. Humanitarian crises. Piracy. Terrorism. Countries on the verge of collapse.

In an appearance before the Senate Foreign Relations Committee in 2009, retired Navy Vice Admiral Lee F. Gunn told lawmakers that "climate change poses a clear and present danger to the United States of America." In addition to the effect diminishing water supplies will have on land operations and rising seas on coastal bases, Gunn observed that while the threats posed by climate change may not be 100 percent certain, a wait-and-see attitude is not an option. "In the military," he said, "by the time threats are 100 percent clear, something bad will have already happened on the battlefield. We can't wait to act."

The concentration of CO2 in the atmosphere currently sits at 392 parts per million (ppm), which is the ratio of the number of carbon dioxide molecules to all the molecules in the atmosphere. That's up from the approximately 275 ppm that the atmosphere absorbed in pre-industrial times—meaning most of the past 650,000 years. Between 2000 and 2007, atmospheric CO2 climbed an average of 2 ppm annually, the fastest seven-year buildup since monitoring began in 1959.

If the global community continues down this same path, climate change experts project that atmospheric CO2 levels could reach 550 ppm by 2050. That's a number that keeps many scientists and national security specialists up at night, since it's expected to result in a calamitous rise in average global temperatures of at least 3 degrees centigrade, or 5 degrees Fahrenheit.

WEAVING THE CO2 BLANKET

Atmospheric CO2

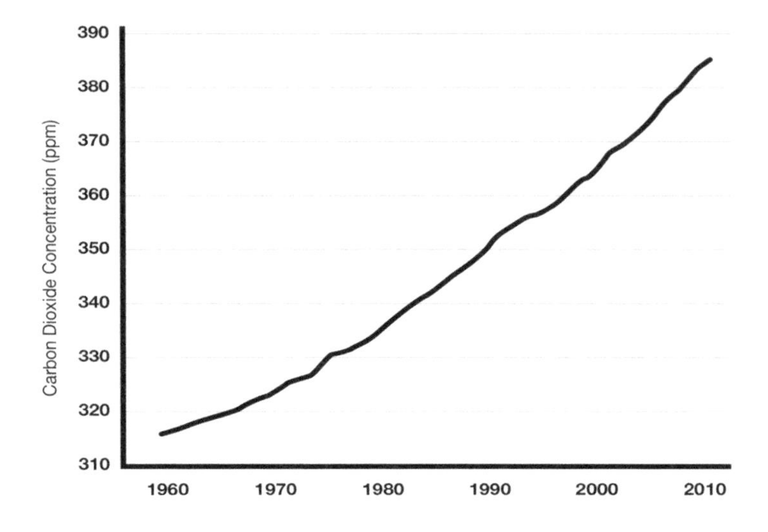

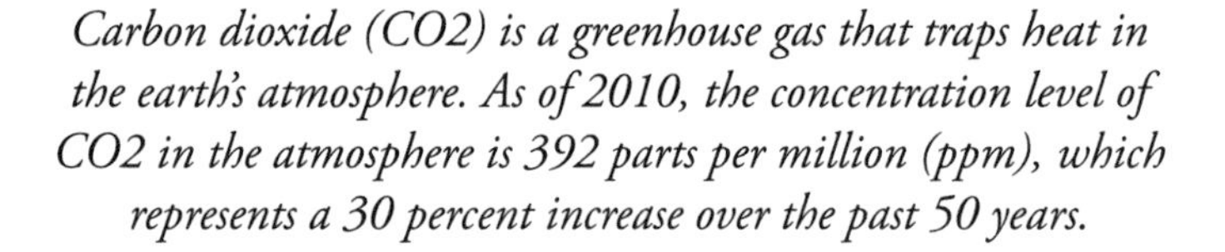

Carbon dioxide (CO2) is a greenhouse gas that traps heat in the earth's atmosphere. As of 2010, the concentration level of CO2 in the atmosphere is 392 parts per million (ppm), which represents a 30 percent increase over the past 50 years.

That's the bad news. The good news is that our situation is far from hopeless. Some CO2 in the atmosphere isn't a bad thing; without CO2 and other greenhouse gases to trap heat in the atmosphere, the planet would be too cold for humans to inhabit. The issue is simply a matter of how much. Climate experts report that if we can get the carbon dioxide in the atmosphere back down to no more than 350 parts per million—the magic number scientists have determined to be the safe upper limit for CO2 emissions—as soon as possible, we'll be in a solid position for avoiding runaway climate change and returning the planet to some form of equilibrium. Since natural carbon sinks such as plants and the ocean allow the earth to rid itself of nearly half the CO2 released into the atmosphere each year, limiting the amount we force it to absorb will give it a chance to recover.

As with a patient who must lose weight or lower his cholesterol, achieving the goal requires changes large and small—everything from installing energy-efficient light bulbs and increasing recycling to pursuing alternative forms of energy and developing ways to capture and store the emissions spewing from factories.

One thing we don't have to do, however, is give up our cars. Not that we ever would, of course. We love our cars. They're an essential component of the American lifestyle. But thanks to EVs, we're off the hook. Although our cars are a big part of the problem today, transitioning to electric vehicles will make them part of the solution tomorrow.

"The scheme of combustion to get power makes me sick to think of it, it's so wasteful."

—Thomas Edison

CHAPTER 6

Electrons vs. Molecules: The Economic Smackdown

Achieving energy independence is great. But it gets even better. The main reason that nearly every car on the road will be electric in 20 years comes down to pure economics: electric miles are much cheaper than gas miles, and it's a cost differential that will only grow over time.

Although the molecules in a tank of gas offer high energy density—meaning you get a lot of bang for your buck in terms of volume—the biggest drawback of transportation powered by an internal combustion engine (ICE) is that it is grossly inefficient. According to the Environmental Protection Agency, only about 10 to 15 percent of the energy from the fuel in the tank actually moves your car down the road—with just 1 percent moving the passenger and the remainder lost to engine and driveline inefficiencies and idling. In fact, idling alone squanders almost 20 percent, which isn't surprising when you consider just how often and how long cars sit in bumper-to-bumper traffic each day, air conditioners humming away.

100 PERCENT IN…10 PERCENT OUT

Inefficiency of an Internal Combustion Engine Car (ICE)

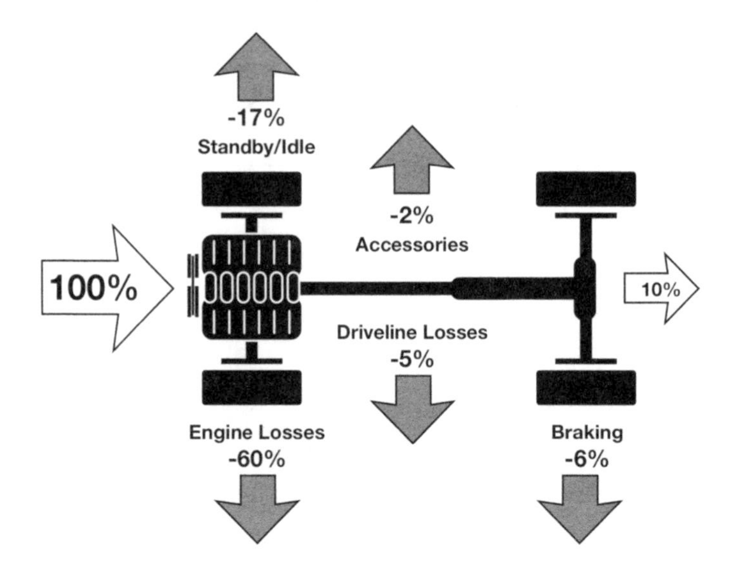

Inefficiencies in the internal combustion engine car result in a loss of as much as 90 percent of the energy in gasoline. Only about 10 percent of the energy from the fuel in the tank actually gets you to your destination.

Using electrons to power a motor is another story. Electric motors transform more than 90 percent of the energy from electricity into mechanical energy.

And here's where it really gets good. In addition to an electric motor's efficiency, the electricity used to run it is significantly cheaper than gas. The U.S. average cost for electricity is 10.2 cents per kilowatt hour. Compare that to $3 for a gallon of gas, and you end up with a pretty big difference. A highly efficient gas-powered car getting 30 miles to the gallon costs around 10 cents per mile; an EV running on electricity costs just 2.5 cents per mile. Similarly, a typical car getting between 20 and 25 miles per gallon travels approximately 8 miles for every $1, while that same $1 takes you 50 miles in an electric car.

HOW DOES 100+ MILES PER GALLON SOUND?

2010: Miles Per Fuel Dollar—EV vs. ICE

50 miles

8 miles

A gas-powered car covers just 8 miles for every $1. That same $1 takes an electric vehicle 50 miles—the equivalent of approximately 150 miles per gallon of gas.

THE EV ADVANTAGE GROWS EVEN WIDER

2015: Projected Miles Per Fuel Dollar—EV vs. ICE

56 miles

7 miles

Over the next 5 years a gas-powered car will become 12.5 percent more expensive to fuel, while an EV will become 12 percent less expensive to fuel—increasing the EV's advantage over an ICE car by another 25 percent.

Auto engineers have done great work improving gas mileage and limiting emissions. But technology isn't able to do a whole lot more to fix the waste inherent in a gas engine—any more than the sailing ships of the mid-nineteenth century were able to stave off the steamship by adding more sails. Although initially requiring sails to boost the limited power of its engines (not unlike a hybrid vehicle, which uses an internal combustion engine to supplement its battery), the steamship had the big advantage of being able to sail when there was no wind. Over time, however, technology improvements saw the steamship power past its rival no matter what the conditions. Many shipbuilders weren't ready to see the writing on the wall, however, and spent the next decades engineering innovations to increase the speed of the sailing ship, most of which entailed ever-taller masts and a greater number of sails. But no matter how many sails were added, the vessels simply could not keep up. Quite the opposite, in fact. Enormous schooners sporting towering masts and oceans of sails eventually grew so top-heavy and unwieldy that they essentially tipped over.

A GAS MOLECULE'S TRAVELS

The Long Road From the Oil Field to the Gas Tank

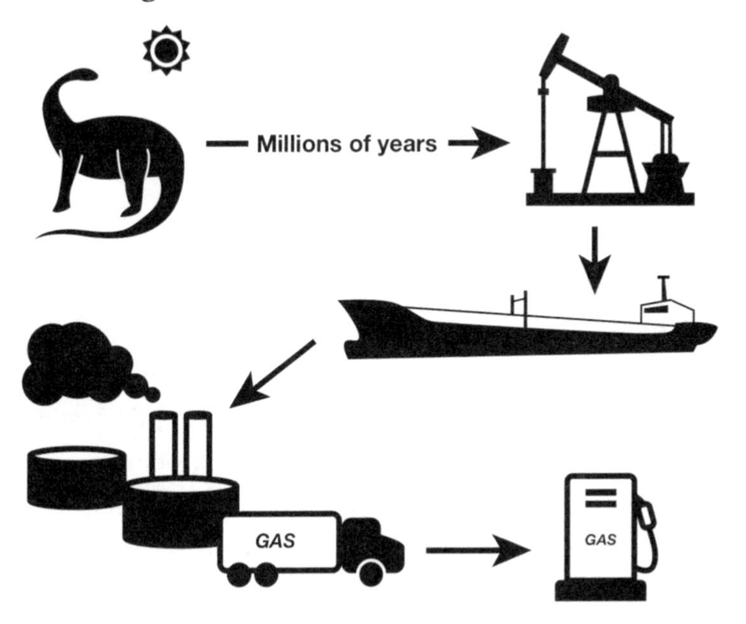

Gas molecules must travel a long distance—including months of transport and refining time—before ever seeing the inside of an automobile fuel tank.

Still, the inefficiency of a gas-powered engine is only part of the equation. There's also the issue of the so-called well-to-wheels comparison, or how far gas molecules must travel before ever seeing the inside of a fuel tank: the eons it took for decomposing, solar-rich plants and animals to sink deep into the earth to become petroleum, only to be pumped back to the surface millions of years later and sent on a long journey via tanker ship and a series of trucks—first to the refinery and then to the gas station—each step along the way requiring substantial output of the very energy being recovered and transported.

And all to end up in an engine that wastes most of what remains.

Domestically produced electricity, on the other hand, is simply pushed through the grid and directly into the car, with little energy loss suffered along the way. Critics argue that since 70 percent of U.S. electricity is generated from fossil fuels, all cars with cords do is move the vehicle's emissions from the tailpipe to the smokestack. But that's not true. Electric cars are always cleaner and more efficient. Even today, with over half of all U.S. power coming from coal, the dirtiest source of electricity available, EVs put out a total of one third fewer emissions than gas-powered cars, simply because they are so much more efficient. Analysis from a joint study conducted by the Electric Power Research Institute and the National Resources Defense Council suggests that plug-in hybrids and EVs could cut CO2 emissions by more than 450 million metric tons annually by 2050. That's the equivalent of removing 82.5 million gas-powered vehicles, or about a third of the cars currently on American roads. The study's authors also point out that the benefits will only multiply as renewable and nuclear sources of energy grow in prevalence, meaning that unlike gasoline cars, plug-ins will only get cleaner as they age.

Electric vehicles are also cheaper to maintain, since they have 70 percent fewer moving parts than an internal combustion engine. There's no engine, no ignition, no catalytic converter, no muffler. No rods, cams valves, pistons, or piston rings. No smog checks, no oil changes. And with less to go wrong under the hood, there's less in the way of service and maintenance, a problem only for owners of gas stations and auto dealerships, which in 2004 saw operating profits of nearly 60 percent from their parts and services departments, according to the now-defunct *AutoExec* magazine.

Even the brakes on EVs rarely need replacing. Take your foot off the accelerator and the regenerative braking system—which turns the electric motor into an electricity generator—gathers energy as the car slows, reusing that energy upon acceleration. Energy expended climbing uphill is actually recovered on the way back down, rather than being wasted burning up brake pads. There's not even enough brake dust to dirty the wheels. Dave Ross, a popular Seattle radio show host and Prius evangelist, boasts on-air that he has driven his car over 100,000 miles and has yet to need a brake job.

ELECTRIC CARS ARE CHEAPER TO MAINTAIN

Gas Car vs. Electric Car—Maintenance Cost Comparison

	Gas (ICE)	Electric (BEV)
Engine / Motor	$400	$50
Transmission	$100	$50
Cooling System	$50	NA
Exhaust	$50	NA
Filters	$25	NA
Brakes	$100	$50
HVAC	$25	$25
TOTAL	**$750**	**$175**

Electric cars have a third the number of moving parts as gas-powered cars. They require no oil changes or tune-ups, minimal transmission repairs, and few brake jobs.

Though the battery itself features plenty of electronics, it has no moving parts to malfunction or break. In fact, recent improvements in lithium-ion battery technology—resulting in increased energy density and decreased weight, all for less cost—have finally enabled the EV to become a viable alternative to the gas-powered engine, with massive headroom for improvement in the coming years.

And although range anxiety, the fear of being caught on the road with a dead battery, is often cited as a reason to stick with a conventional car, it's important to understand that in the U.S. an estimated 90 percent of all vehicle trips are less than 30 miles, including work commutes; more than half, in fact, are less than 6 miles. According to the U.S. Bureau of Transportation Statistics, the average car is driven just 32 miles per day, well within the 100-mile range of many of the electric vehicles coming on the market.

EVs AND THE AMERICAN COMMUTE: A PERFECT FIT

Miles Driven Per Vehicle Trip

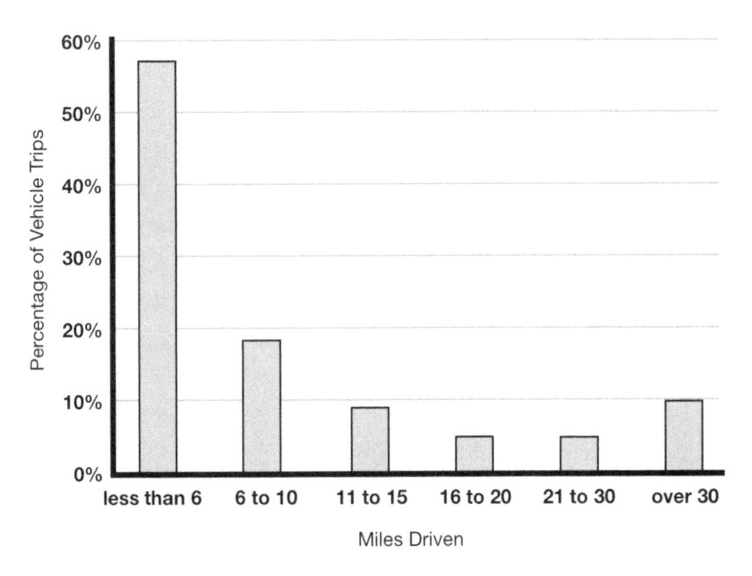

Ninety percent of American cars are driven less than 30 miles each day, with more than half of all U.S. vehicle trips under 6 miles. Such travel patterns are well within the electric-only driving range of the coming plug-in hybrid vehicles (PHEVs) and full-battery electric vehicles (EVs).

Most all-battery EVs, when fully charged, have less than a third the range of a typical gas-powered car, which makes people understandably nervous when home is the only consistently predictable fueling location. But just as driving a conventional car at a quarter tank keeps a driver on alert, so too does driving an electric car. (Looked at another way, a 20-mile range is the new quarter tank.)

Even so, long-time EV owners report that they quickly became confident in their routines, given that almost all American vehicle trips fall into the under-30-mile range. Drivers with a 20-mile commute each way—40 miles total—still have plenty of room for unexpected detours to stop by the store or pick up dry cleaning.

That said, anyone too uncomfortable with the EV's 100-mile range should instead buy a plug-in hybrid. PHEVs feature the same cost savings as EVs for the first 40 miles, when the cars operate on the battery. After that, drivers rely on the backup gas engine for any extended-range driving needs.

In any case, don't let the 100-mile range fool you. Except for the silent motor, these cars are nothing like the "golf cart" EVs of yesteryear. Today's electric cars are unbelievably fun to drive, featuring superior handling, instant acceleration, and 100 percent torque right out of the blocks—not to mention a dashboard filled with enough graphic displays to satisfy any auto lover's inner geek.

In fact, an EV's green credentials are increasingly only part of its appeal—so much so that Tesla doesn't spill all that much ink highlighting the environmental angle when marketing its high-performance all-electric Roadster. Sure, it's there, and buyers appreciate that the car burns zero gas. But what all of them really like is that the Tesla goes fast, very fast—from 0 to 60 in under 4 seconds. That's Porsche Turbo territory. And all for just 2 cents a mile.

WHY ELECTRIC CARS ARE SO MUCH FUN TO DRIVE

Torque Comparison—Gas Engine vs. Electric Motor

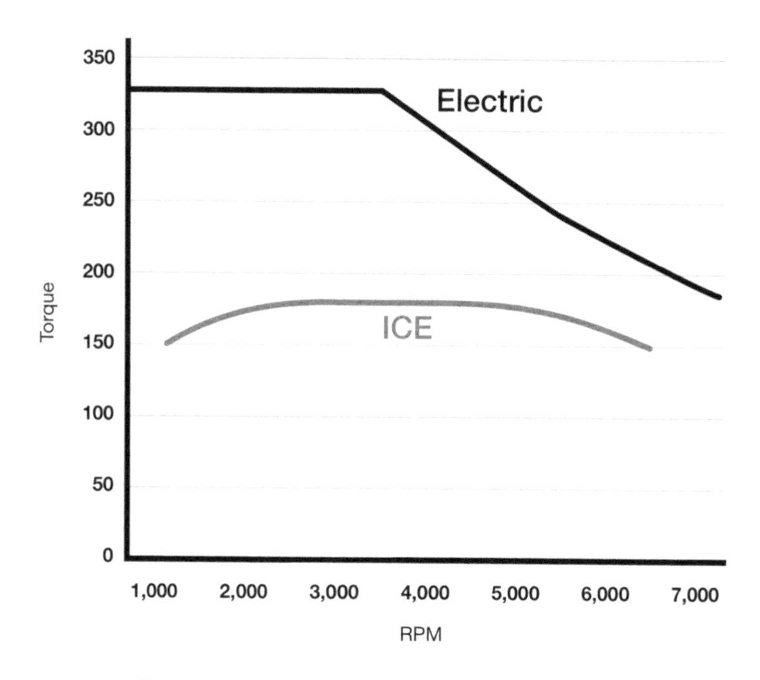

Unlike gas-powered cars, EVs have maximum torque from the get-go, which means they can accelerate incredibly quickly. The Tesla Roadster, which has an electric motor the size of a rugby ball, goes from 0 to 60 in under 4 seconds—putting it right up there with a Porsche.

No matter how you look at it, paying 2 cents a mile is good for the pocketbook. But it's also beneficial when considered from a national perspective. The United States is currently home to some 250 million of the world's cars. The Electrification Coalition predicts that at least 14 million cars cruising U.S. roads will be electric by 2020, and that EVs could account for three quarters of all personal transportation miles by

2040. If so, U.S. oil consumption in the light-duty fleet would plunge to just 2 million barrels a day, from nearly 9 million, ending the need for oil imports from unfriendly countries.

In the end, it all comes down to drivers becoming comfortable with the new technology. Americans have been driving internal combustion engine vehicles for over a hundred years. ICE cars are familiar, and we like what's familiar—even if it isn't always the best. Yet over that 100-year period, automobile drivetrains have evolved and changed, and today we have electric starters, V-8 engines, and automatic transmissions, to name just a few modifications.

So consider this: an EV is really nothing more than the next-generation drivetrain, with the seating, driving interface—steering, signaling, braking, accelerating, etc.—and accoutrements all largely the same. And just as past changes in drivetrain technology have improved the automobile, so it is with the "different" EV drivetrain.

But in this case, different is also better. For example:

- EVs are incredibly quiet...Rolls-Royce has spent millions trying to achieve the same decibel level in an internal combustion car.

- EVs have instant torque and are therefore able to accelerate really quickly...Porsche has spent millions to achieve the same 0-to-60 results.

- EVs are stretch-your-legs-out roomy, thanks to electric motors the size of a rugby ball, one-speed transmissions, and batteries that can be sandwiched below the seats and in other spare spaces... Mercedes Benz has spent millions to optimize passenger and cargo space.

Then there's the fact that electricity scares people. Once again, however, the idea of electricity under the hood is only frightening because it's unfamiliar. Which it shouldn't be, since conventional cars already have batteries under the hood, batteries that are used to start the engine and provide extra power for lighting and other accessories.

What's really odd, though, is that the idea of carrying a tank of gasoline with 20 times the energy density of a battery pack doesn't seem to bother anyone. An Internet search of information regarding gas tank accident statistics shows all the top hits to be legal organizations offering to represent victims or the relatives of victims who have been injured or killed in gas tank explosions and fires. Which means there are more than just a few incidents each year. (In fact, there are about 250,000 a year.) And what's even more amazing is that more than 30 years after Ford Pinto gas tanks exploded following collisions, there still are no national safety standards that regulate the design of fuel tanks and systems.

It's also worth noting the potential dangers associated with pumping gasoline into your car. Next time you're at a gas station, take a moment to read the long list of things you shouldn't do during the fueling process—one of which includes talking on a cell phone, since static discharge can and does start gas station fires.

Electric car batteries, on the other hand, are extremely safe. Many years of research and laboratory testing have gone into ensuring that the new lithium-ion batteries can tolerate temperature extremes, are well sealed and well protected, and automatically disconnect in the event of an accident. In any case, the Toyota Prius, which has been on the road for more than 10 years, has a battery system similar to that found in the newest EVs, but you'd be hard-pressed to find a case of anyone having been injured by a battery (though no doubt it would really hurt if you dropped one on your foot).

Americans currently spend an average of 15 hours—approximately 2 business days a year—driving to gas stations and pumping gas. (And of course the gas gauge needle is always close to empty when you have the least time to gas up.) Taking into account oil changes, brake jobs, and other forms of maintenance, your car's internal combustion engine extracts up to three days from your life each and every year. Compare all that to sticking a simple plug into a socket at the end of the workday when you return home—the place you're headed anyway.

Once again, when it comes to EVs, different really is better.

Coming Down The Road

Car	Type	Maker	When	Why
GM-Volt	PHEV 5-Door Hatchback	GM (U.S.)	Late 2010	Bet-the-company product with U.S. government backing. Huge marketing push.
Prius	PHEV 5-Door Hatchback	Toyota (Japan)	Late 2010	Toyota has the most experience building hybrids. Strategic for No. 1 carmaker.
LEAF	EV 5-Door Hatchback	Nissan (Japan)	Late 2010	Renault-Nissan largest single commitment to EVs
Focus	EV 3-D Hatchback	Ford (U.S.)	Late 2011	Mainstream and cheap. Ford will not be left behind.
Blue Zero	EV, PHEV	Mercedes Benz (Germany)	Late 2011	Platform approach promises broad range; economically built
Fluence	EV Coupe	Renault (France)	2011	Synergy with Nissan models
Model S	EV 4-Door Sedan	Tesla (U.S.)	2011	Affordable luxury. Its $57,000 price tag seems a bargain compared to the $110,000 Roadster. Claimed range of up to 245 miles.
E6	EV 4-Door Wagon	BYD (China)	Available in U.S. 2011	Warren Buffet invested $230 million
Coda	EV 4-Door Sedan	Designed in U.S., manufactured in China	Fall 2010	Tag line: "Invented in America, Built Globally." Range of 120 miles. U.S. headquarters in Los Angeles.

Vehicle Systems—From ICE to EV

Internal Combustion Engine Vehicle (ICE): A conventional vehicle that stores gasoline (or diesel) in a fuel tank. That fuel is burned in the engine, which delivers mechanical energy to drive the car. Because of the high energy density of the gas stored in an on-board tank, ICE vehicles are able to travel several hundred miles before needing to be refueled. Gas-powered vehicles are highly inefficient, however, wasting more than 85 percent of their energy.

Hybrid-Electric Vehicle (HEV): Like a conventional ICE vehicle, an HEV has an internal combustion engine and requires gas. Additional energy is stored in a battery, which powers an electric motor; that motor converts electric energy into mechanical energy. The two technologies work together to power the car, resulting in reduced fuel consumption and tailpipe emissions, and making the vehicle more efficient and cleaner than one powered by gasoline alone. A hybrid has no plug; the batteries can only be charged when the vehicle is running. A combination of gas and electric power gives hybrids the same or even greater range than conventional cars.

Plug-In Hybrid Vehicle (PHEV): A PHEV is essentially a regular hybrid with a bigger battery and a smaller combustion engine, and can be plugged into a standard 120-volt or 240-volt outlet. PHEV batteries can propel the vehicle in all-electric mode at modest speeds for 7 to 60 miles, depending on the size of the battery pack, before requiring assistance from the internal combustion engine. Once the battery pack is depleted, the gas engine is engaged to drive a generator to power the electric motor and extend the range of the vehicle. PHEVs get up to 100 miles per gallon. In a parallel PHEV such as the plug-in Prius, the wheels can be driven either by the gas engine, the electric motor, or both. In a serial PHEV such as the GM-Volt, the wheels are driven only by the electric motor.

Electric Vehicle (EV): The EV has no internal combustion engine, instead relying on electricity stored in its battery pack to move the car. EVs produce no emissions. Recharging the battery entails plugging into a standard wall socket or charging device that receives power from the grid. The car can also be hooked up to another electrical source such as a solar panel. The newest EVs coming on the market have a range of about 100 miles, while the Tesla Roadster has a claimed range of up to 245 miles. (Also known as a Battery Electric Vehicle, or BEV.)

"The United States was not built by those who waited and rested and wished to look behind them. This country was conquered by those who moved forward."

—JOHN F. KENNEDY

CHAPTER 7

Trading the Oil Barrel For the Watt Bucket

So won't all those electric vehicles greedily sucking power from the grid cause it to collapse?

Not even close. In a normal day of electricity production in the U.S., somewhere between 40 and 50 gigawatts of power are being produced at any one time on the national grid. That's an enormous amount of power. Despite the occasional blip due to extreme demand, storm-related outages, or human error, the system performs at nearly 100 percent on a day-to-day basis.

National power demand drops way off late at night and then ramps back up in the morning as we turn on our lights, coffee pots, and computers, and our factories kick into gear. But the U.S. power sector is designed as an on-demand system, meaning an overcapacity of power is always available whether we draw on it or not. Massive power plants can't be turned on and off with the push of a button. They take a long time to shut down, and they come back up again very slowly. So any time of the day or night, we can flip a switch and the lights come on, the air conditioner blows, the computer beeps to life. As a result, the late-night hours see a huge gap between what's being produced and what's actually being used, with all excess electricity essentially run into the ground and wasted, since the grid currently has no way to store excess power.

THE WATT BUCKET

***U.S. Daily Electricity Production vs. Usage—
Billions of Watts (Gigawatts)***

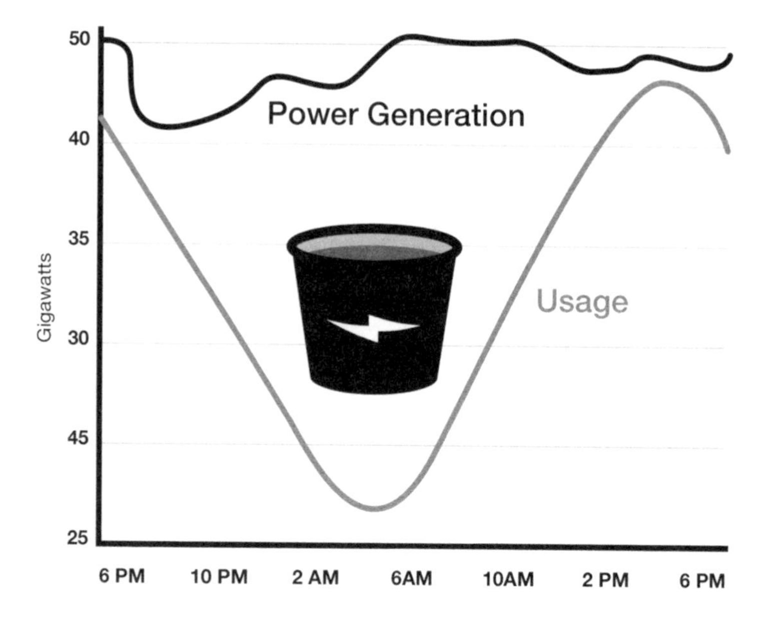

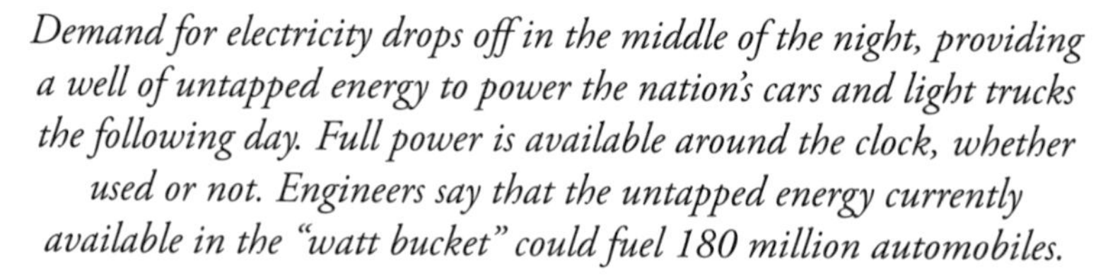
Demand for electricity drops off in the middle of the night, providing a well of untapped energy to power the nation's cars and light trucks the following day. Full power is available around the clock, whether used or not. Engineers say that the untapped energy currently available in the "watt bucket" could fuel 180 million automobiles.

But that excess capacity available in the middle of the night —a well of potential energy filling the watt bucket—could be used to charge electric vehicles with little effect on the system as it is today. With absolutely no changes to the grid, we could flow millions of kilowatt hours of energy into the batteries of tens of millions of electric vehicles and propel Americans through 1 billion driving miles the very next day. The U.S. Department of Energy reports that the nation's existing electrical grid has enough off-peak capacity to charge 180 million EVs without adding any new capacity in the form of coal or nuclear power plants. In addition, a Pacific Northwest National Laboratory study shows that the existing grid is fully capable of supporting 84 percent of the nation's cars and light trucks, which could potentially eliminate 7.25 million barrels of oil a day—completely ending our need to import oil from OPEC nations.

GRID OVERLOAD—THE GREAT MYTH

10 Million EVs—Impact on the Existing Grid

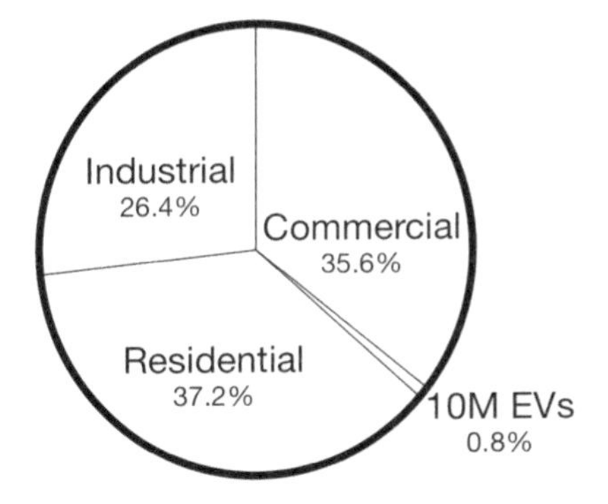

Ten million electric cars drawing electricity from the grid would have less than a 1 percent impact on the current system. The existing grid also has enough off-peak capacity to handle 180 million EVs without utilities needing to generate more electricity.

Plug-in hybrids use about as much electricity in a year of standard driving as a three-person household's water heater consumes in five months, according to the Idaho National Laboratory. Energy officials note that while all-electric cars use approximately four times more electricity than a plasma TV, the fact that the utilities have successfully dealt with the increased power demand from mass consumer adoption of those TVs assures them that a similar rollout of EVs won't cause a problem—particularly since the majority of the cars will be charged at night. The utilities know what's coming, and they'll have time to adapt. In addition, energy companies can help distribute the load on the grid by charging higher rates during peak-demand hours and lower rates at night.

WHICH USES MORE ENERGY? AN EV OR A REFRIGERATOR

Energy Use Comparison—Typical Electric Appliance vs. Electric Car

Annual Energy Usage-Electrical Appliances (KWH)	
Home Heating System	3,524
Central Air Conditioning	2,796
Refrigerator/Freezer	2,610
Water Heater	2,552
Electric Car (40 mile commute)	2,500
Clothes Dryer	1,079
Lighting	940

In a single year, a typical refrigerator consumes more electricity than an electric car driven 5 days a week for a 40-mile round-trip commute.

Even assuming significant adoption of new electric cars over the next 10 years, given that Americans buy between 12 and 14 million new cars each year, it will obviously be quite a while before the number of EVs on the road represent the majority of the total U.S. fleet of 250 million vehicles. (The Department of Energy reports that in 2008 the median lifespan for a car was 9.4 years, and 7.5 years for a light truck.) During that time, however, the electrical grid will not only become cleaner as utilities turn to wind, solar, and nuclear energy—all made in the U.S.—but also more efficient as the existing infrastructure is expanded and updated. While some of that expansion and modernization will result from the implementation of a national energy policy—and the federal funding that fuels it—the majority will occur organically among the nation's more than 3,000 utility companies in response to increased customer demand for plug-in vehicles. Capitalism is a powerful force in America and one of our greatest strengths.

The electrical grid in its existing form is an amazing achievement, reliably delivering power to the nation's industries, its commercial buildings, and over 120 million households. Nonetheless, an upgraded grid is key to the future of electricity as the nation's primary energy source for cars and light trucks. Rather than remaining a simple grid that merely distributes power, the so-called smart grid of tomorrow—currently under construction thanks to some of the best engineering minds in the country—will allow consumers to generate and store their own power using micro-generators, solar energy, and fully charged EVs. Some of that power can then be sent back to the grid during periods of high demand. The result will be better alignment of supply and demand, greater efficiency, and less need for additional power plants. A smart system will also have the capacity to monitor and adjust nonessential use as needed—automatically triggering a car to charge its battery or a dishwasher to start, but only during less expensive, off-peak hours. As long as the electric vehicle's charger is programmed to have the car's battery ready at a specific time, it doesn't really matter when during the night that charging takes place.

Finally, one of the best things about turning to electricity to power the personal transportation sector is that it is comes from diverse sources, including coal, natural gas, solar, hydroelectric, geothermal heat, nuclear, and even landfill gas. (Just 1 percent of U.S. electric power is generated from petroleum.) What's even better from an energy security standpoint is that all these sources can be generated in our own country. With so many different options available, any interruption in the delivery of one—a lull in the wind, a delay at the coal plant—is easily made up elsewhere.

ELECTRICITY: MADE IN THE USA

U.S. Electricity Production by Source

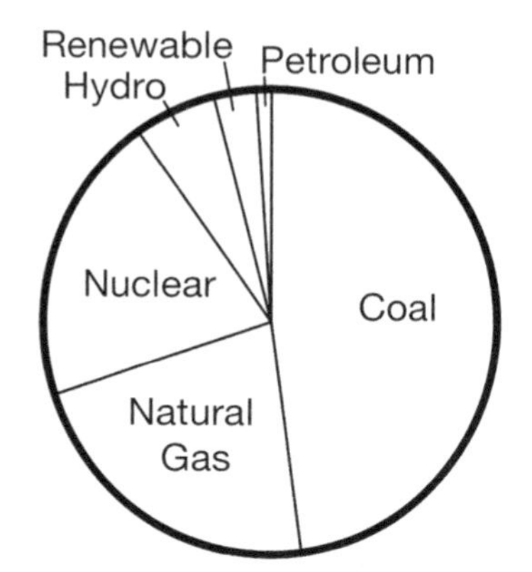

Electricity is generated from a variety of U.S.-based sources. While the majority of American electricity is currently generated from fossil fuels such as coal or natural gas, the fastest growing source of electricity is wind power.

Unlike oil, electricity prices are stable, since they're far less subject to the uncertainties and instabilities that cause volatile price swings. Even though electricity prices tend to fluctuate on a day-to-day—even hour-to-hour—basis during high-use periods in summer, for example, they have remained relatively consistent over the long term. U.S. Department of

Energy numbers show that over the past 25 years, the average retail price of electricity rose around 2 percent per year, and that prices climbed by more than 5 percent per year only 3 times during that period. In addition, although fossil fuel prices jumped 21 percent between 2004 and 2006, the average price of residential electricity increased from 9 cents per kilowatt hour to just a bit over 10 cents.

While price stability owes much to the fact that utilities offer inexpensive off-peak pricing, the recent infusion of government stimulus money into the development of renewable energy sources has gone a long way toward keeping prices low. Gradual adoption of energy-efficient light bulbs and household appliances will also do much to keep both electricity demand and cost down. It can even be argued that because a utility's capital infrastructure and its workforce are already in place, increased demand for electricity could even result in lower prices. While no one should ever bet on a government-regulated industry to drop its prices, historical data and business trends strongly indicate that electricity prices will at the very least remain stable.

ELECTRICITY COSTS ARE STABLE

Electricity Prices Since 1980

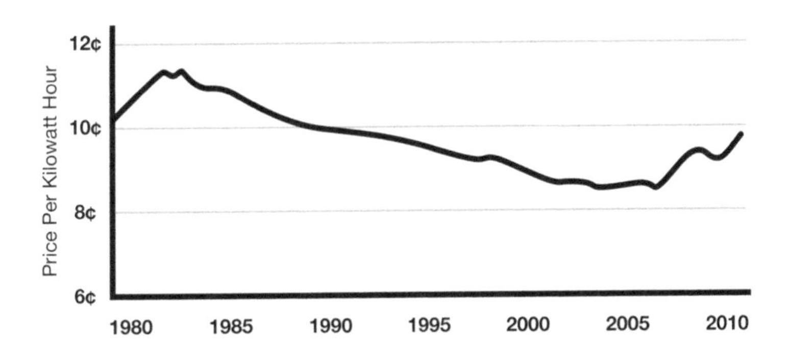

Unlike oil, electricity prices have remained—and will continue to remain—stable over the long term.

Far from causing the electrical grid to collapse, an EV-based personal transportation system will actually encourage it to grow and diversify. Even more, once the infrastructure to support them is in place, electric vehicles will enable us to continue our uniquely American love affair with the automobile, which along with the nation's highway system, is the backbone of our culture and economy. With economics, security, and greenhouse gases taken out of the equation, there remains no good reason, beyond personal choice, to junk our cars in favor of public transportation. And given how inadequate public transit is across so much of America, that's a very good thing.

CHARGED UP

All EVs coming on the market have a built-in charging system that can be plugged directly into any household power outlet. The time needed to charge the battery will vary depending upon how much energy is in the battery at the outset of the charging session. A completely drained EV battery using standard 110-volt household power can take up to eight hours to fully charge. Many people, however, will choose to install Level II (240-volt) chargers in their homes; such chargers can be purchased along with the electric car. The equivalent of an electric-dryer circuit, Level II chargers will cut the maximum refueling time to between two and three hours. In addition, because most of these charging systems are "smart," they can be programmed to deliver electricity to the car during off-peak hours, when demand is low and the grid has significant excess capacity.

Even though up to 80 percent of vehicle charging will take place at the vehicle owner's residence, many states and cities have plans and funding in place for a rollout of Level II charging stations throughout public roadways and parking areas. At least a half a dozen U.S. companies are already in the business of designing and manufacturing charging stations to meet the growing demand. As electric vehicles become more common and the public charging infrastructure expands, Level II charging will become increasingly available at work and in public areas, allowing drivers to top up their batteries while they eat or shop. Quick-charge Level III chargers, capable of charging an EV in just a bit more time than it takes to fill today's gas cars at a gas station, will also become more widely accessible, and will bring a battery that has been drained back up to 80 percent charged in under 20 minutes.

---

*"There are no constraints on the human mind,
no walls around the human spirit, no barriers to
our progress except those we ourselves erect."*

—Ronald Reagan

CHAPTER 8

ZERO TO SIXTY: THE EV AS THE ECONOMIC TURBOCHARGER

As if achieving energy security, saving money at the fuel pump, and limiting emissions weren't incentive enough to embrace electrons over molecules, there is yet another reason to make the switch.

The electriconomy.

While dependence on oil to drive the U.S. economy is a financial drag on both the individual and the nation, the electrification of passenger cars represents a huge chunk of the worldwide alternative energies market. As stated earlier, venture capitalist John Doerr predicts that the clean tech market will become a $6 trillion annual economic boon. That's six times the size of the Internet and computer markets *combined.* Experts estimate that alternative energy industries, which include the electric car and its associated ecosystems, could support up to 20 million jobs worldwide in the energy sector alone over the next two decades. And it is that huge chunk—the electrification of our cars and light trucks—that will lead future American economic growth.

Worldwide, there are about 1 billion vehicles on the road today, and industry experts suggest the planet could see as many as 2 billion cars within 20 years as private car sales in China, India, and other developing nations skyrocket. Vehicle sales in China, for example, jumped 46 percent to almost 14 million in 2009 as that nation overtook the U.S. to become the world's largest auto market, something it wasn't expected to do until 2016. And given China's eagerness to develop its EV market and infrastructure, an increasingly large percentage of the cars hitting Chinese roads over the next two decades will be electric. In fact, China recently unveiled a revitalization plan for the domestic automobile, calling for production of half a million full-electric and plug-in hybrid-electric vehicles by 2011. Much more than China's domestic market is at stake, however. Exports of batteries, electric vehicles, and the associated infrastructure now offer China an opportunity not only to maintain global manufacturing dominance but also to become an innovation leader.

Though it will be a long time before it leaves oil and coal behind, China is choking on its own development and has opted to take the greener road, proving once again that necessity is the mother of invention. Alarmed by its growing reliance on foreign oil, as well as by the threat posed to public health and its economy by the pollution filling its air and clogging its waterways, China has made the development of alternative energy technologies and the electrification of its transportation a national priority. In fact, the Chinese government recently issued a mandate that the country lead the EV domain—and the clean energy production equipment industry as a whole—within three to five years. And since official decrees are easily carried out in countries unrestricted by public participation in the political process or independent ownership of key utilities, we can count on it happening.

According to U.S. Energy Secretary Steven Chu, China has been pouring $100 billion a year into alternative energy investments, more than five times the U.S. total of $18.6 billion for 2009. When you consider that Chinese wages are less than half those of Americans, that means China is applying over 10 times the resources to its EV and alternative energy efforts compared to the U.S.

With vast numbers of newly prosperous citizens purchasing private cars—Beijing alone sees some 1,500 new cars added daily, with an anticipated 5.5 million vehicles expected to clog that city's streets by 2015—the Chinese government is quickly stepping up its investment in the EV sector, offering subsidies for EV and hybrid purchases and issuing grants to encourage innovation in battery development. In addition to mandating that the state-run electricity grid install electric vehicle charging stations in the nation's 13 largest cities, China recently announced a new program to subsidize plug-in cars in five major Chinese cities over the next two years; all-electric car buyers will receive up to $8,800, which is more than a year's salary for the average Chinese city worker.

China is already the largest EV manufacturer in the world, with vehicle production expected to exceed 10 million units this year. "The fuel-efficient and new energy vehicles should account for 10 percent of the total industry in 2012," said China's Science and Technology Minister Wang Gang last year. According to the *Wall Street Journal*, the June 2010 decision to launch the two-year subsidy program "reflects the government's determination to foster an electric-car industry capable of competing head-on with global auto makers that are rushing to launch plug-in hybrid cars and all-electric vehicles over the next few years."

An aggressive EV strategy makes sense for China, since reducing oil dependence and shrinking its carbon footprint—approximately half its pollution comes from the transportation sector—are both part of a national strategy. In addition, as *The Green Market* pointed out in a blog, "EVs represent an opportunity for China to leapfrog past the internal combustion engine the same way it was able to skip the massive investment in hardwired phones through the introduction of cell phone technology."

China, however, wants to do more than just compete. A recent study released by the U.S.-based National Foreign Trade Council (NFTC), a free trade advocate, examines policies put in place by the Chinese government to promote the development of its renewable energy sector and shut out foreign participation. The study details a series of Chinese government measures that have stimulated demand for Chinese-made renewable energy equipment, including:

- Preferential financing
- VAT rebates
- Tax incentives
- Procurement preferences for Chinese-owned and Chinese-controlled companies
- Local content preferences
- R&D subsidies for renewable energy equipment producers

In commenting on the study, EERE Network News, a newsletter on energy efficiency and renewable energy distributed by the U.S. Department of Energy, notes that "China's Renewable Energy Law—enacted in 2006 and strengthened in 2009—requires utilities to buy all available renewable power and pay full price for it, while offering it at a discount to their customers."

EERE also highlights the fact that the share of wind power equipment imported into China has fallen steadily from about 75 percent in 2004 to about 25 percent in 2008. "In 2009, Chinese imports of U.S.-made wind turbines fell to zero, after reaching about $15 million worth of imports in 2008," it wrote. "Meanwhile, China has rapidly expanded its solar cell production, nearly all of which is exported."

CHINESE HANDCUFFS

China Discourages Imports of Alternative Energy Technologies

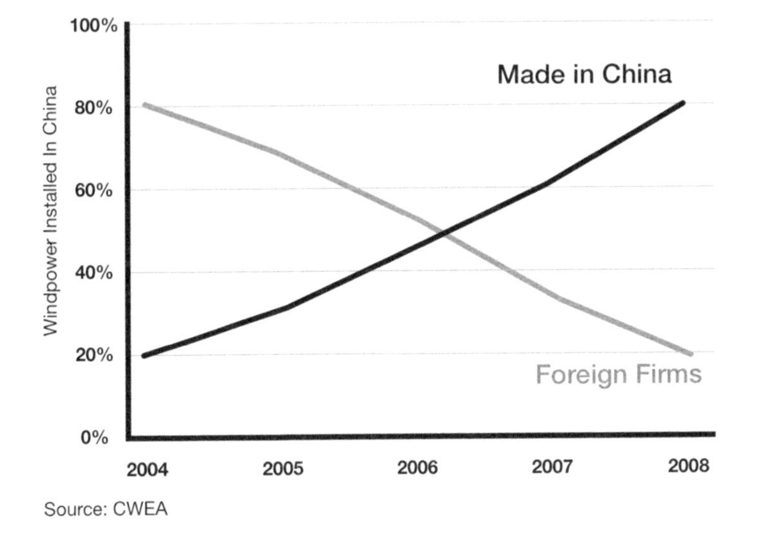

Through government-backed incentives and other nationalistic mechanisms, China is systematically reducing the purchase of imported alternative energy products in favor of Chinese-built goods.

As the NFTC study points out, "Chinese firms stand to gain substantially" from the measures outlined above. And when it comes to the electric vehicle industry, the stakes could not be higher. Whoever captures a significant portion of the EV market is looking at an impressive market share. In 2008, Warren Buffet purchased 10 percent of BYD, a little-known Chinese company, for $230 million. Better known for its cell phone batteries than its cars, which it has only been building since 2003, BYD came out of nowhere to zoom past its bigger, better-known rivals in the race to develop affordable electric cars. BYD's plug-in, available only in China for now, travels over 60 miles on a single charge and costs just $22,000, significantly less than the Prius and GM-Volt plug-ins scheduled to enter the U.S. market at the end of the year. While BYD's model lacks many of the luxury, convenience, and safety features Americans expect from even their most basic cars, the all-out effort put into the vehicle is indicative of just how serious China is about the plug-in market.

But electric vehicles themselves are only part of the expanding economic picture. Widespread adoption of EVs will create demand for charging stations along highways and in cities and the home, a market that is still in its infancy. Battery technology is another area with substantial room for development, since the ideal electric car requires storage devices that are smaller, cheaper, and more efficient than today's technology allows. And although studies show that the electrical grid is currently equipped to sustain millions of electric cars, upgrading the national network will facilitate the development of the smart grid. This high-tech upgrade will enable homes, businesses, and EVs to communicate with utility companies, thus optimizing energy use.

Future scenarios also call for a system that will allow energy stored in EV batteries to be fed back into power grid as needed during times of high demand, which will enable greater use of renewable energy. The good news is that many of these associated industries will generate hundreds of thousands—ultimately millions—of jobs for high tech and "green collar" workers, who cannot be outsourced to other nations.

Unless the U.S. moves aggressively to capture the export-friendly aspects of the EV market and its associated industries, however, it is likely to be little more than a second-tier player in the new economy, particularly when it comes to manufacturing. The lithium-ion battery is a telling example. Although breakthroughs in battery technology have mostly come from the U.S. over the past few years (as have numerous other technology innovations), almost all advanced battery manufacturing has moved to Asia to accommodate the portable electronics industry, which is based predominantly in Japan, South Korea, and China. But because battery manufacturing is heavily automated, labor costs are less of an issue; in fact, the fixed costs are similar no matter where the factory is located.

And because battery packs for cars are heavy, shipping costs become an additional and significant factor.

In short, there's no reason that the U.S. can't jump-start battery manufacturing closer to home—and plenty of reasons why it should. While the worldwide market for the types of batteries that power EVs amounted to just $900 million in 2009, Deutsche Bank estimates that the global market for such large-scale lithium-ion batteries will climb to $15 billion by 2015. (By way of comparison, the market for smaller lithium batteries used to run laptops and cell phones is currently estimated at $7 billion annually.) In an acknowledgment of the huge potential of the domestic battery market, both General Motors and Ford have moved aggressively to site their newest battery manufacturing plants—and the jobs they create—in the U.S.

With a significant portion of the estimated $6 trillion annual alternative energy industry tied up with EVs and their technologies, the U.S. simply cannot afford to be left behind when it comes to the development and manufacturing of batteries—or any other component of the plug-in market. It's also a mistake to assume that there's no point attempting to reinvigorate U.S.-based manufacturing, that China can do anything we can do, only more cheaply. For all its advantages in terms of cheap labor and limited barriers to entry, China is no unstoppable economic juggernaut. Although the expense of doing business in China is generally still less than in the United States, costs there are beginning to rise as that nation's increasingly skilled labor force, finding itself in high demand, calls for more pay and better work conditions.

Still, China often lacks the required skill sets, something Chen Bin, director of China's National Development and Reform Commission's Department of Industry, acknowledged last year. "Many [Chinese] companies simply do not have the research and development capabilities," he said, adding that those with insufficient command of core technologies must acquire key components elsewhere.

At the same time, many U.S. companies have entered China with high hopes of success, only to be thwarted by government regulations, policies that mandate joint ventures with local companies, and seriously inadequate protection of intellectual property. And as evidenced by the recent conflicts between the Chinese government and Google, China doesn't hesitate to retaliate against any company, foreign or otherwise, that pushes back against its soft-power/hard-power approach. It was certainly no accident that after its decision to abandon the Chinese market rather than submit to censorship, Google found its deal with China's biggest cellular company canceled.

But Google is hardly the only foreign firm to feel the sting of China's nationalistic tactics. A March 2010 survey released by the American Chamber of Commerce in China showed that 38 percent of foreign firms operating in China said they felt increasingly unwelcome, up 12 percentage points from just a few months earlier. In fact, the trend was even more pronounced among technology companies, with nearly 60 percent reporting that government policies were detrimental to their businesses. The Chamber attributed much of the increased pessimism to China's "indigenous innovation" policy, outlined above, which encourages both local and foreign companies to choose domestically developed technology, even if inferior, over foreign options.

Regardless of China's policies, however, there is no question that the United States is uniquely qualified to drive the EV revolution. In addition to the fact that many of the breakthroughs in EV technology currently being implemented overseas originated in the U.S., America has a long and innovative history as a technology front-runner—a history stretching from Boeing to biotech, from microprocessors to the Internet. The United States is home to world-renowned universities and research organizations. It has an experienced investor community looking to back promising technologies and talent. It has a mature and robust free market, and courts dedicated to the protection of individual and property rights. Last but certainly not least, the U.S. boasts a productive, entrepreneurial workforce eager for jobs in a new and promising industry.

"I don't want to just reduce our dependence on foreign oil and then end up dependent on foreign innovations," President Obama told a crowd in economically depressed Indiana after announcing $2.4 billion in government grants to stimulate the manufacture and development of EVs and next-generation automobile batteries. "I want the cars of the future and the technologies that power them to be developed and deployed right here."

Electric vehicles are coming. American consumers will buy them. That much is fact. But who will create and build them? Who will benefit most from the electriconomy? Now is the time for the U.S. to step up and take the lead in their development and manufacture. With the future of the American economy at stake, the cost of doing nothing is simply too high.

"Taking a new step...is what people fear most."

—Fyodor Dostoevsky

CHAPTER 9

So Who Wants to Buy a Texaco Station?

While there will be winners and losers along the EV highway, the net effect—as with every other technology shift, from steamships displacing sailing ships to the global impact of the Internet revolution—will be a giant positive as the EV-driven economy takes off. Texaco Stations and Jiffy Lubes will gradually be replaced with Level II and Level III charging stations, many located right outside your workplace and favorite restaurant. And since many of the jobs that will support the electriconomy can't be moved offshore, EVs and their associated ecosystems offer clear economic advantages in countless arenas, many yet to be developed. No one knows exactly what all these industries will look like, but there's no question they'll provide countless opportunities for people with the guts and vision to tackle them.

The key, however, is making sure you're on the winning side of the EV equation. Smart owners of horse-drawn buggy shops sent their sons to learn the auto mechanic trade in 1910 instead than having them apprentice at home. Savvy leather workers focused on the dimensions of Henry Ford's Model T seats rather than on those of the new buggy maker down the road. Similarly, if you own a family muffler shop today, now might be a good time to steer the next generation into some aspect of the up-and-coming electric car ecosystem.

As new car sales shift to electric vehicles, the population of cars with an internal combustion engine will continue to age, providing ongoing opportunities to service those vehicles. Over time, however, the need for traditional auto mechanics and service providers will shrink as all remaining gas-powered cars are replaced by plug-ins. The catch? Those jobs won't be replaced by an equivalent number of electric vehicle mechanics and service providers. EVs don't need mufflers, lube jobs, or smog checks, and they require far fewer brake jobs and motor repairs. Nonetheless, the Jiffy Lubes of the world will need to supplement their shops with the equipment and know-how to diagnose and service electric drivetrains and battery packs.

Similarly, while the new cars won't need gasoline, they will need electricity. As drivers become more comfortable with the new technology and the charging infrastructure grows to meet demand, they'll increasingly look for places to top up their batteries while away from home—and they won't be pulling up to the pump at any of today's 157,000 gas stations. In addition to upgrading their service shops and training traditional auto mechanics to work on electric vehicles, forward-looking gas station owners should plan to invest in quick-charge Level III charging stations in anticipation of the day when most drivers are looking for electricity rather than gasoline.

Just as cars always need refueling, people need refueling, too, which is why fast food franchises and coffee shops along major highways should consider installing charging units in their parking lots. The same is true of big-box retailers, shopping mall owners, cineplexes, and office complexes. Drivers looking to top off a battery while they stop for lunch, shop for new pair of shoes, or take in a movie represent a phenomenal business opportunity.

In the initial stages of electric vehicle adoption, owners of publicly available charging stations might choose to absorb the cost of customer charging, since the cost will be low and the public relations benefits high. Later, however, as EV ownership grows, businesses might want to charge a small convenience fee for the use of each charging station. Companies willing to make the capital outlay to provide charging stations across a town, city, or state could deploy "subscription models," just as the wireless phone industry does today.

As we move to an electriconomy, electricians and technicians will be needed to install and maintain home and public charging stations and networks, producing millions of new jobs. With electric vehicles, all homeowners have access to a personal fueling station—comparable to the 500-gallon fuel tank, complete with a gas pump, buried beneath the driveway of a very wealthy Seattle homeowner. EV owners who wish to upgrade from a 120-volt to a 240-volt system to accommodate a Level II home charging unit—equivalent to installing a washer/dryer unit—will need a licensed electrician to install the wiring.

In addition, each electrified public parking spot needs a charging station, and there's a lot of money to be made installing and maintaining that hardware. EV entrepreneurs might also want to look into the possibility of contracting with municipalities for public right-of-way charging stations. The setup is likely to be similar to agreements currently made with cable television providers, whereby Comcast and other cable companies given monopoly rights in certain areas are required to install and maintain an infrastructure to support the system. Finally, with the smart grid still in its infancy, the nation's 3,200 electric utilities are going to need additional staff to maintain the new technology and infrastructure.

Hardware, however, represents only one aspect of the EV charging equation. Since data about a vehicle's charging status will be linked wirelessly to the utility company, an industry focused on monitoring and coordinating those communications is already under development, encompassing everything from devising payment options to remote access of charging information. Nissan and Ford are already working to create a smartphone application that will allow EV owners to check their battery's status remotely. And while the Nissan LEAF is the first car that will be a full participant in the smart grid, it will by no means be the last. Smartphones, smart cars, smart houses—all will be connected to one another and to the smart grid of the future.

The important point is that whether accessed through cars, cell phones, or home computers, the smart grid will dramatically change the (inter)face of transportation, allowing drivers to schedule off-peak charging, access their payment plans, warm or cool their cars remotely, and determine a vehicle's charging status. Other systems will enable utilities to draw power from fully charged vehicles during times of high demand, transmit directions to the nearest charging station, or provide data on a car's remaining charge, which will allow drivers to modify their acceleration as needed. In the end, building out the electrical grid networks might well become a bigger driver of economic growth than the cars themselves.

Still, there is plenty of room for growth when it comes to the cars. The future of EV technology lies in the race to build the lightest, cheapest, and most powerful battery. The winners of this race are sure to find eager automakers lining up outside their doors, since automakers are engaged in intense development with suppliers—some well known, others up and coming—to increase energy density and decrease costs. The same is true for developers and suppliers of integrated components of the EV powertrain, where breakthroughs are happening on a regular basis.

Looking beyond the cars themselves, the EV revolution also opens the door to reintroducing large-scale manufacturing jobs in America. This is a particularly important point since it is crucial from both an economic and national security perspective that electric cars, their batteries, their components, and the associated infrastructure be made in America. And while manufacturing has taken a beating here, it is not yet dead and buried. In fact, manufacturing is making a comeback, and EVs are a big part of that resurgence. Need proof? Just consider Tesla's recent decision to partner with Toyota to manufacture its cars at Toyota's shuttered plant in Northern California. Or Nissan's new Tennessee plant, where it will produce battery packs for the LEAF. Or GM's new factory in Michigan, where it assembles batteries for the Volt—the nation's first such plant owned by an American car company.

Those batteries also offer unprecedented recycling opportunities. Unlike gasoline, which once combusted in a car's engine is gone forever, lithium-ion batteries are fully recyclable. A lithium-ion battery no longer able to store enough energy to power a car 100 miles still has years of charging cycles in it, and is fully capable of a second life as a storage unit for windmills and other stationary objects. Once a still-useful battery has been pulled from a car, a broker can sell it to a power utility. Second-life batteries can be used by businesses and individual households, or placed on wind farms to store energy captured during off-peak hours for later use. And since sustainability, particularly in crucial industries, is a hot issue today, it's no surprise that Carlos Ghosn, CEO of Renault-Nissan, has spoken publicly about his company's commitment to recycling EV batteries.

But there's more. Widespread adoption of EVs and the EV lifestyle will mean overarching changes for our society, changes that stretch far beyond issues of transportation and employment, and into the realm of urban planning. As solar panels, windmills, and other methods of producing electricity see significant expansion in the coming decades, homeowners will be able to produce their own electricity. Those concerned with their carbon footprint, not to mention their ever-increasing energy bills, can run their cars and homes with their own personal power plants—generating millions of jobs as electricians and other technicians are called on to build and install them. People with big energy-guzzling houses in the suburbs or country will no longer have any reason beyond personal choice to give them up and move closer to cities or public transit. When it comes to solar power, at least, our famously oversized McMansions are ideal power-generators. Goodbye greenhouse emissions and goodbye guilt.

Last but certainly not least, distributing power generation all the way to the home offers Americans greater national security. While it is conceivable that terrorists could attack and disable centralized power plants, it is difficult, if not impossible, to target 110 million homes.

PART II

Out in Front

"The best way to predict the future is to invent it."

—Alan Kay, former Apple Computer guru

The electriconomy is right around the corner, and plug-in vehicles will soon become a way of life. And although the notion of refueling a car by plugging into a wall socket might still seem as strange as the idea of searching the Internet for movie listings did just 15 years ago, it won't be long before charging a car takes no more thought or effort than charging a cell phone.

As with the Internet revolution before it, however, the EV revolution has been a long time coming. Some incredibly smart people have been working many years to make it happen. Whether designing the next-generation battery or constructing policy that will become a model for the nation, the individuals featured in the following section have dedicated themselves to inventing and advancing the future. They have invested their time—and often their money—creating and enabling products that will change the world for the better.

Each is an expert in a particular aspect of the electric vehicle ecosystem, from automobiles to energy to infrastructure. Each is recognized for innovative thinking and for being ahead of the curve. And each is known for the ability to make things happen, sometimes even when no one was interested in what he or she had to say. Some are driven by a desire to see the U.S. (or, in one case, China) achieve energy independence. Others seek promising business opportunities. Still others care passionately about the environment. What they all share, however, is a common vision: the electrification of our personal transportation system.

Since the concerns of this book are predominantly American, nearly all the people covered in the next section represent American companies. But because another, equally important component of the book considers what the U.S. needs to do to win the EV race, the final profile is of a phenomenally successful Chinese entrepreneur who has his eye firmly fixed on the plug-in prize. So much so that he was able to entice Warren Buffet—the Oracle of Omaha, Mr. Middle America himself—to enter the EV waters by investing in his company first.

CHAPTER 10

LIFE IN THE FAST LANE

BOB LUTZ, GENERAL MOTORS

Calling Bob Lutz an automobile expert is like calling Ferrari a car. Not only has Lutz achieved legendary status throughout the auto industry, he's the ultimate car guy's car guy. After nearly 50 years in the business, Lutz officially retired from his role as GM adviser earlier this year, closing–but not slamming–the door on his life as an auto executive that includes stints as vice president of Ford Motor Company, president of Chrysler, executive vice president of sales at BMW, and vice-chairman at General Motors. On the eve of his departure from GM, Lutz discusses why he considers the Volt his most important legacy, explains his call for higher gas taxes, and reaffirms his belief that global warming has absolutely nothing to do with human activity.

Asked which car was the most significant of his career—the Opel GT? the Viper?—Bob Lutz, the larger-than-life retiring General Motors executive, pointed to the Volt, GM's extended-range EV, due out at the end of 2010.

"I'm an acknowledged global warming skeptic, but I was, in fact, the progenitor of the Chevrolet Volt—the earliest proponent at GM of a modern-day electric vehicle," he said. "I took a lot of abuse for it because of our unfortunate financial experience with the EV1," the all-electric vehicle GM released in 1997 and pulled the plug on just two years later. "That was then compounded into an unfortunate PR experience, where we did exactly the wrong thing when it came to canceling peoples' leases. Canceling the leases was bad enough. But then crushing the vehicles gave rise to the whole conspiracy theory. I got so many emails from people saying, 'You rotten son of a bitch, I hope you rot in hell, with the hundreds of millions that you got from the oil companies.'"

The 78-year-old former Marine Corps pilot—Lutz still flies his jet fighter plane, his King Air twin-turboprop plane, and his helicopter regularly—circled back to his views on the human effect on climate change, which in the past he has vociferously denounced as "a crock of [expletive]." "I am an anthropogenic global warming skeptic, and proud to be," said the man known to many as "Maximum Bob," his gravel-scraped voice filling the room at the sprawling GM Technical Center in Warren, Mich. Anthropogenic? "It means that temperature changes on the planet are caused by human activity. Which they're not. We're all in favor of a clean environment, but I simply and honestly do not believe that SUVs are melting the planet. If we've got to get rid of SUVs, it's because they use a lot of a precious commodity called oil, which is going to become more and more expensive, and which geopolitically speaking is not in the best of hands. So I see a certain geopolitical vulnerability to the nation unless we can get ourselves off petroleum. That's what motivates me."

In January 2010, in fact, Lutz came out in favor of a gradual increase in gasoline taxes as a means of weaning the nation off its insatiable appetite for oil. Though he received a public drubbing for even suggesting it—something that doesn't appear to faze him much—Lutz is still an advocate of the tax. "Personally, yes, though I'm reluctant to state it publically, because of all the hate mail I get along the lines of, 'That's fine for you, you overpaid corporate swine, but those of us who commute to a job that just barely pays for the fuel, if gas goes up another 50 cents, I don't know how I'm going to get to work.' That's true, but that merely tells us that we have structured the transportation system in this country wrong. And that instead of like in Europe where everything is densely packed and more vertical, in the U.S. we have this infatuation with the idea that, we're America, everybody deserves their own half-acre with a ranch house on it."

Pointing to the sprawling metropolitan regions across the U.S. and the long driving commutes Americans tolerate for the sake of affordable housing, Lutz blamed it all on inexpensive gasoline. "If you look at the macro picture—the excesses in this country, the waste—it's all driven by cheap fuel." But the country can no longer tolerate a rapid doubling of fuel prices, he said, as happened in the summer of 2008 when gas went from under $2 to over $4 a gallon. "You cannot manage a product portfolio to deal with discontinuous change like that. If I were divine emperor of the U.S.—the only political job I'd accept—I would ordain a 20 cent per gallon increase in the fuel tax every year for the next 10 years. That would do two things: It would gradually get people used to higher gas prices. And it would permit people to plan their next purchase. They would say, 'Gas is $3 now, Sweetie, but in five years it's going to be $4. So do we really want to buy this?' It would drive consumer behavior."

The electrification of motor vehicles is a given, Lutz said, although he acknowledged that the rate of adoption is dependent on what happens with battery technology, and he predicts the transition is likely to be a gradual one. "First of all, lithium-ion battery capacity has to build along with the volume of the cars," he said. "We've got to get better range; the cost of all this stuff has got to come down. If we can double the capacity of lithium-ion, EVs become a lot more viable, even if a small internal combustion engine continues to accompany the vehicle as a backup power plant."

What fills Lutz with pride, he said, is that the Volt is a complete departure from today's technology. "I was explaining the Volt to a friend in Montserrat, and he said, 'That's like my three-masted sailboat, with a diesel engine on board.' I said, 'Bingo! Great metaphor. Wind dies, fire up the diesel, and get home.' The Volt is an entirely new and different thing, and it's going to be hugely successful."

But the fact that it is an entirely new and different thing initially caused problems with GM's engineers, who kept trying to make it into a very efficient hybrid, with a gas engine available to boost the car on an uphill, for example, rather than an electric vehicle with an internal combustion engine backup that would take over if the battery were drained. "I said, 'No! This vehicle is not designed to be the world's most efficient series hybrid,'" he recounted. "'This vehicle is designed to give 80 percent of Americans—whose daily trips are 40 miles or less—completely fuel-free driving.' And the engineers said, 'Yes, but it's more efficient if you occasionally run the gas engine.' And I said, 'I don't CARE!' I mean, that's when I decided that our own people—the engineers—didn't understand the concept, so I finally said, 'Look, guys, think of it this way: It's not a parallel hybrid. It's not a series hybrid. Think of it as an electric vehicle with a 40-mile range. Do you get that?' 'Yes, we get that.' 'Okay, and now add an emergency generator to get you home if the battery's down.' And that's the way I like to look at it. It's an electric car, with extended range."

After that, Lutz said, everything was smooth sailing. "Quite the opposite of what Toyota predicted," he added, referring to the Japanese car company's dismissal of lithium-ion's readiness as a battery option. "They confidently stated that we would fail." Reminded of his response that Toyota would eat those words, Lutz grinned. "Yeah, they didn't digest them well."

He added that the Volt will do more than just get a driver home, thanks to the 12 gallons of gas on board. "As we do more and more research around the world and you present people with the various concepts—EV with quick charge, EV with battery swap, or EV with less range but with a range-extending internal combustion engine like the Volt—83 percent opt for the Volt solution," Lutz said, noting that the numbers on the West Coast were particularly high. "Based on the intelligence we're getting, almost every car company in the world is going to do a Volt-like vehicle."

Lutz said he was tired of the media painting a coming showdown between GM and Toyota. "It just drives me nuts that some in the general press and news magazines hype 'The Coming Technology Battle: Toyota's Plug-in Prius vs. Chevrolet's Plug-in Volt,'" he said. "The Prius is going to have something like six or seven miles of electric range on a good day." Based on hundreds of tests in all terrains and weather conditions, Lutz said the Volt will get about five times as much on a single charge. "On a cold day with aggressive driving, you're going to be down to 30-something [miles of EV range]," he said. "But on a warm day with mild driving, we've had people get 50 miles."

Queried on how GM planned to generate excitement around the Volt, Lutz indicated the car would take care of that itself by attracting several different types of early customers. "First there are going to be early adopters for profound environmental religious reasons," he said. "Those are the same Hollywood people who were the early adopters of the Prius, and the same people where we had to pry the EV1s out of their hands. So the West Coast Hollywood community is guaranteed. Then every politician in the country will want one, to demonstrate green credentials. But then there will be guys like me who just love exciting automotive technology. The thing is phenomenally fun to drive—a great chassis, wonderful handling and steering—it's a very engaging car. Acceleration is terrific. And everything takes place in this eerie silence. There's no gear noise or electric motor whine. It's just this totally silent rush of power, which is very addictive."

Lutz didn't hesitate when asked how he'd like to be remembered. "I'd like to be known as the guy who was always a proponent of automotive excellence, regardless of which company I worked for," he said. "I would also like to be known as the person who got GM refocused on product excellence. As I say, if you look at an automobile company as a society, the product is the aristocracy of that society, and design is the king."

CHAPTER 11

Success is a Team Sport

Mike Tinskey, Ford Motor Company, and **David Cole,** Center For Automotive Research

Two automobile industry thought leaders outline why the economics behind electric vehicles mean the cars are finally here for good, as well as the reasons the EV revolution will be consumer-driven. In calling for a resurgence of American-based manufacturing, they also argue that Americans must get their hands dirty if the U.S. is to achieve national success and security.

Best known for his catch phrase, "Once you plug the vehicle into the wall, success becomes a team sport," Mike Tinskey, manager of sustainability and electrification at Ford Motor Company in Dearborn, Mich., spends his days focused on facilitating the marriage of the automobile and the electrical grid. What intrigues him most is the idea of electrification of transportation as an enabler for greater good. "I would say the most exciting thing for me isn't necessarily the driving part of transportation," he said. "It's what it does for our economic security in terms of oil imports. And what it's doing for us in terms of our smart grid." Electrification of the nation's transportation system in a vacuum is not necessarily the right thing to do, the best thing to do, he said. "But when you combine it with all the other, more strategic, big-picture items—including economic and energy security, protecting the environment, and utility infrastructure modernization—it can be a great enabler."

David Cole, chairman of the Center for Automotive Research, based in nearby Ann Arbor, shares Tinskey's enthusiasm for an updated electrical grid and greater national security, but warns that that any marriage between the next-generation automobile and the electrical grid requires some government involvement. "Government policies need to be sufficiently agile to track the technology and the economics," he said. "We've got a lot of electric capacity if we charge overnight. But if we charge between 4 p.m. and 6 p.m. we're in trouble. There have got to be economic incentives not to do that. In an emergency, people should be able to charge but there's got to be a serious economic penalty for that. With a smart grid, though, we can handle all that."

Cole stressed, however, that government involvement only works if the economics of plug-ins make sense. Forcing EV adoption through mandates isn't effective. "Ultimately," he said, "the final judge on this technology is not going to be a politician. It's going to be the consumer. If we push things too hard and the economics don't make sense, there's a risk of a big backlash." Tinskey concurred, noting that the success rate of a regulation-driven program "does not always meet market needs."

Both men, however, believe that advances in battery technology mean the economics for EVs and hybrids make sense this time around. The proof? All the major car companies have one or the other or both coming out in the very near future. In fact, Ford is planning to build a full range of models for the next-generation Focus, with hybrid and electric-only powertrains, all on the same assembly line and on the same shift. Ford is also electrifying a variety of its existing models for the global market, a strategy in line with its belief that at this point no one technology works for all customers.

It may take some time before plug-ins see widespread adoption, but Tinskey and Cole agreed that plug-ins are here for good. Asked what barriers remained to their widespread adoption, Cole answered that there were none. "Everybody is full-bore on the new technology." Tinskey said that the fact that lithium-ion batteries have now reached the point where they are able to carry a medium-sized car approximately 100 miles on a single charge represents a crucial threshold. "In terms of energy, density, and size, I think we're right on the knee of the curve, where there are enough sales to generate the R&D necessary to keep the improvement curve going," he said.

Tinskey also thinks the U.S. is at a crucial juncture in terms of the plug-in adoption rate. "If you look at the whole spectrum—starting with traditional hybrid vehicles like the Prius and the Ford Escape, which just recapture regenerative energy, all the way up to pure EVs—we're probably at the flash point right now," he said. "My view is that you will see this niche market—the one or two percent of the market, the early adopters with disposable income—who go after pure battery electric, and that will happen relatively soon. And then as the battery gets better and better, you'll see mass adoption." Tinskey estimated that electric cars with a plug—pure EV or hybrid—will make up a quarter of all new Ford sales by 2020.

It's the second-generation lithium-ion batteries that will make all the difference, Cole believes. He predicted a cost reduction of more than half, down to about $250 kilowatts per hour. "A $4,000 or $5,000 battery at the 40-mile range like you find in the Volt is probably a critical point," he said. "Once you get to that point, then the game changes. Does that occur in 2 years, 5 years, or 15 years? We don't know. But once it hits that particular area, things are going to get very interesting."

Both agreed that a return to manufacturing in the U.S. is crucial, and not just to the success of the plug-in industry. "The important thing from a U.S. perspective is not where the technology comes from, but to have employment," Cole said. "You want to be producing the key [components] here, like batteries and motors. People say Asia has this huge head start with lithium batteries, but it doesn't mean anything because that's for cell phones, computers, and power tools. What we are looking at is a whole new industry, with a whole different set of characteristics in the battery."

He recalled a conversation he'd had a year earlier with the head of the auto division of Japan's Ministry of Industry and Trade. "One of his comments was that every industrialized economy in the world understands the role that manufacturing plays in their economy—with the U.S. as the one exception," Cole said. "We felt we could do without it, but it's fundamental. Tech is great, finance is great, but you can't eat it, wear it, live in it, or drive it to work. Ultimately, if an economy is going to be successful, you have to get your hands dirty, you have to make things. And when you really get down to it, one of the most fundamental parts of our economy that has had the tar beaten out of it is manufacturing."

Battery technology offers the U.S. an opportunity to catch up in manufacturing, Tinskey said, particularly since the size and cost of plug-in batteries mean it isn't all that cost efficient to import them from overseas. "At 400 to 500 pounds, they just aren't conducive to shipping long distances," he said. "But with all the government funding available now, we have a chance to turn that around. The research is already out there. Now we just need the manufacturing."

CHAPTER 12

SMART MONEY

RAY LANE, KLEINER PERKINS CAUFIELD & BYERS

The EV expert at the world's most recognized and successful venture capital firm explains why electric vehicles and their associated technologies represent an unprecedented business opportunity. A flag-waving American who favors energy independence, Lane details the importance of establishing a national energy policy, outlines why Americans must stop buying oil from countries that despise the U.S., and describes what role the government needs to take in establishing plug-in technologies.

Ray Lane figured driving a prototype Fisker Karma to visit a potential investor would only help his cause. But when he was pulled over by a policeman while on his way to the home of Larry Ellison, CEO of Oracle, Lane realized driving the unregistered premium plug-in hybrid sports sedan might not have been the best idea. "'What is this thing?'" he recalled the officer asking. "I had to explain that the car wasn't street legal. But he didn't give me a ticket. He loved the car, and wanted his picture taken with it."

As the partner overseeing EV investments at Kleiner Perkins Caufield & Byers, a Silicon Valley venture capital firm with an impressive track record as an early investor in more than 300 information technology and biotech companies over the past 35 years, including Netscape, Amazon,

Google, Sun Microsystems, and Genentech, Lane knows a thing or two about the benefits of electric vehicles. But his main reasons for promoting them aren't what most people might think.

Although linked by profession and friendship to some of the more visible names associated with progressive causes, including fellow Kleiner Perkins partners John Doerr and Al Gore, Lane is quick to dispel any notion that his principal rationale for backing EVs is environmental. Seeing a reduction in emissions would be great, said the former president and CEO of Oracle, but it's just one reason to fund electric, and his primary motives lie elsewhere. First and foremost: it's a great business opportunity. "There is a lot of money to be made in EVs," he said. "You could ignore the whole environmental angle—the Al Gore argument of why we ought to be investing in clean tech—and still have a damn good business. EVs will not save the planet in the next five years, but they could be an awfully good investment."

The other predominant—and more personal—motivation concerns the problem of U.S. reliance on foreign oil. "I'm a flag-waving American and I believe in this country," Lane said. "But we don't have an energy policy. We're idiots when it comes to an energy policy. The U.S. is simply borrowing money to burn it."

While Lane believes the majority of Americans won't buy EVs on environmental grounds unless the price is right, he thinks concerns around national security could swing them. "If we can get the price of an EV close enough to that of a car with an internal combustion engine, then I think consumers will say, 'Okay, let's stop buying fuel from—and being dependent on—people who don't like us. I'll pay a few thousand dollars more because I love my country.' That's why I love T. Boone Pickens. It just makes a lot of sense."

The problem, of course, is that EVs can't compete financially with the internal combustion engine right now without relying on subsidies to bring down costs (just as subsidies to the oil companies keep gas prices artificially low). Lane, who believes that GM's extended-range hybrid Volt is the best long-term technology given issues around range anxiety, doesn't see EVs able to hold their own until batteries come down in price to $200 per kilowatt hour. Current costs for low manufacturing volumes range from $500 to $600 per kilowatt hour.

In light of the national security issues at stake, however, he believes the government needs to play a big, even aggressive, role in launching plug-ins as an industry. "Right now we're reliant upon a single fuel source for our transportation," he said. "Electricity is a multiple fuel source—solar, wind, gas, coal. That argument alone should have the U.S. government putting together all sorts of incentives and policies, doing whatever it can to wean us from oil."

He gave a rueful chuckle, aware how odd it sounded for a self-described Republican, moderate or otherwise, to call for subsidies. "If I weren't in this business I'd probably be on the Tea Party bus," he joked. "Instead, I'm standing in front of the Department of Energy asking for handouts." Then, growing more serious, he continued: "Government is just a pipeline to spend our money. So I care about what they spend our money on, and this is one area I think they should spend our tax dollars."

Another reason to get the ball rolling in terms of subsidies and credits, he said, is that the EV industry is going to be a highly competitive field, one that the U.S. can't afford to miss out on. While Lane's view is that from a national security perspective it's not particularly relevant where the EV's components originate, he believes it is critical from an American competitiveness standpoint. "This is a huge growth industry and we need to fight for jobs and leadership," he said. "The U.S. represents one of the largest auto markets in the world. And the whole fleet will change out over the next 10 years. The more components that come from here, the more jobs we create."

Looking down the road, Lane sees China as the main competition, particularly when it comes to battery manufacturing. China already has a great deal of experience; within 10 years, he said, China will produce a high quality car with a good electric drivetrain, just as other Asian countries have come from seemingly nowhere to dominate the auto industry in the past. "Japan did it. Korea did it. Now there is every reason to believe China will do it."

Recalling the difficult time American auto companies had when Honda and Toyota first landed on U.S. shores—how in the 1980s they tried to copy Japan's magic formula for producing better cars at a lower cost—Lane anticipates a repeat with China. "But I don't understand why we can't invent it," he said. "I think we just need to create a new sheet of paper to work from. The way we design cars, the way we build cars, the way we sell cars—change everything! There's more money to be saved in the system than there is in the car itself."

With over 50 investments totaling approximately $1.5 billion, Kleiner Perkins holds one of the largest clean tech portfolios in the industry, and has no intention of scaling back. "We'll continue to invest at this level or even more in the future," Lane said, noting that the sector represents a $6 trillion industry worldwide. "It's going to have a long, long run."

And as a sector it's already off to a blazing start. Lane estimates that 2008 saw $3.6 billion in venture capital invested in clean tech in the U.S. alone, with another $4 billion invested in 2009. That's nearly as much as was poured into each of the more traditional venture capital sectors such as life sciences, software, and semiconductors—industries that have had some 30 years to develop. "It's what I call a perfect storm," Lane said of six-times-faster growth rate of alternative energy technologies. "The national security argument, the greenhouse gas emissions argument, the new technologies argument—all these things have come together to make clean tech very, very cost competitive with older technologies."

Lane is particularly bullish on the business potential of energy storage. "Storage is huge," he said. "It could be the biggest green tech sector of all. If we can figure out how to store electrons cheaply, that's the way we save burning a lot of fuel."

CHAPTER 13

The Electric Fuel Tank

Bart Riley & David Vieau,

A123 Systems

Energy's status as an essential commodity makes it imperative that the U.S. establish energy independence, say two top innovators at one of America's leading battery development companies. They describe how breakthroughs in battery technologies will speed the electric vehicle revolution, why America must lead the way, and what the U.S. needs to do to win the EV race.

As a co-founder of A123 Systems, a hotshot developer and manufacturer of high-power lithium-ion batteries, Bart Riley often speaks to enterprising young graduates eager for career guidance. He gives them same advice he received at a 1990 commencement speech at Cornell University, where he received his M.S. and Ph.D. "The guy told us there were two secrets to life," Riley said, referring to then-president Frank Rhodes. "'First, figure out what you love to do. Second, find somebody to pay you to do it.'" Now the chief technology officer and vice president of research and development for A123,

Riley is doing exactly that. "In the end, it's about passion, and it's amazing what passionate people can do," he said.

Passion, combined with a whole lot of smarts, has taken A123 a long way in a short time. Founded in 2001 with a $100,000 grant from the U.S. Department of Energy, the company's initial public offering raised nearly $380 million in September 2009. The Watertown, Mass., firm was also awarded another Department of Energy grant that same year—this time for the somewhat larger sum of $249 million—as part of a government-backed plan to boost electric vehicle technology.

Using technology developed at MIT, A123 first hit pay dirt with a high-power, long-life battery capable of remaining stable at high temperatures. The problem with earlier lithium-ion batteries technologies, the ones currently used in cell phones and laptops, is that they occasionally catch fire—an occurrence known in the industry as a "thermal incident." That possibility has long deterred their use in electric vehicles and other power applications.

The first company to adopt A123's lithium-phosphate-based chemistry was Black & Decker, which began using it for its hand held tools in 2005. A123 then expanded its focus to include batteries capable of powering electric vehicles safely. The result was a next-generation lithium-ion battery that crossed a significant threshold in terms of cost, weight, and safety. The company recently signed deals with Fisker Automotive, a new American company building premium plug-in hybrids, and Chrysler, which wanted a U.S.-based supplier, to provide their battery systems.

Most experts peg the current cost of an electric vehicle battery at around $600 per kilowatt hour—a number that drops to approximately $400 if produced at high volume. A123 Chief Executive Officer David Vieau estimates that when the price falls to $350, which he expects to see around 2016, the market for EVs and plug-in hybrids will open up significantly. The company, however, hopes to be ahead of the game long before then; according to Riley, A123 is working on some new innovations that could reduce the cost by 50 percent.

"A lot is being worked on within lithium and beyond lithium," Vieau said. "There's a bit of a gold rush occurring—a research renaissance in the area of energy storage." Given that batteries have been improving at a rate of between 9 and 12 percent annually in recent years, Vieau, who labels himself a "technology optimist," anticipates advances resulting in a lightweight battery capable of providing a 300-mile range on a single charge by 2020, which would match the range of a gas-powered vehicle. He's also a self-proclaimed optimist when it comes to the EV market, projecting that plug-ins could represent 20 percent of new-car sales within that 10-year period.

The research renaissance Vieau refers to extends beyond batteries and into the world of ultracapacitors, which store electrical energy by capturing ions, according to Joel Schindall, a professor of electrical engineering at MIT. Because there is no chemical reaction, ultracapacitors are very resilient—they can be charged and discharged very rapidly—and they last a very long time, over a million charge/discharge cycles. Because of these properties, ultracapacitors are often referred to as "mechanical batteries." The problem is that current ultracapacitors store only about 5 percent as much energy as a comparably sized lithium-ion chemical battery. Schindall, however, has found a way to increase the surface storage area and thus increase the energy storage to perhaps 25 percent that of a battery.

"Ultracapacitors are most attractive for short-term energy storage—regenerative braking or engine stop-start so as to avoid idling," Schindall said. "They're also helpful when coupled with batteries so that the ultracapacitor provides the surge current and the battery provides the long-distance operation." Although adding an ultracapacitor to an EV might initially increase the cost, he noted that the flip side is that it should significantly increase the performance of the car in terms of acceleration as well as battery life. That is particularly important given the percentage of the total car cost that plug-in batteries represent.

Whatever the eventual makeup of an EVs storage system, thanks to technology improvements, less toxic materials, careful placement within the vehicle, and an electrical system designed to disconnect when damaged, Vieau is confident that any battery poses significantly less of a threat in the event of a collision than a car filled with 20 gallons of gas. Pointing out that there are approximately a quarter million car fires in the U.S. each year, he noted that the auto industry is not without safety issues. "The key is to have a battery that behaves better than a tank of gas," he said. "We've been able to do that."

He dismissed the oft-stated objection that in adopting lithium battery technology the U.S. is simply trading a limited supply of oil for a limited supply of lithium, much of which is mined overseas. "We don't buy into the lithium shortage story," he said. "There's actually a plentiful reserve. If you were to increase the population of vehicles on the earth several times over the next 30 years and turn all of them into EVs, you would have three or four times the amount of lithium on the planet to serve them." Add the fact that lithium is 95 percent recyclable, he said, and it's clear that supply will never be a problem.

Once the economics of the EV are comparable to that of a gas-powered car, or at least close, Vieau and Riley agree that performance—including superior handling, acceleration, and torque—will push EVs to the forefront. When first faced with a plug-in, Vieau said, an auto executive's reaction is "denial and outright objection." But put him in an electric car and have him drive around the track? "He's a changed person when he gets out. It's hard to deny performance capability." Add reduced maintenance and how easy it is to plug in at night, he said, and EVs will be hard to beat. "No one would have ever imagined that people would spend the amount of money they do to get a more advanced music box," he said of the iPod. "But the experience justifies it, and the EV is going to be like that. I don't know anyone who looks forward to going to a gas station to fuel up. No one talks about the joy of getting an oil change."

The biggest barrier to mass adoption that Vieau sees is uncertainty around ongoing government support. "Right now we're riding a wave of strong government action and support," he said, referring to the $2.4 billion in Department of Energy grants announced in 2009 to stimulate the EV market. "So long as we maintain that and move to the next level, I'm very optimistic about the market. But in the past we've seen the government back off on these things. OPEC has a lot of reasons to keep the price of fuel down."

In addition, he noted that energy, unlike a flat-screen TV, is an essential commodity in its own right. "Our economy, our lifestyles are dependent on energy," he said. "We're willing to go to war over it." Looked at from that perspective, Vieau said, it's easy to understand why government intervention is required to stimulate and support the electrification of transportation. "The evidence of the last 40 years of buying oil—all the problems it has caused are a matter of record," he said. "We need to solve this problem and eliminate our dependence on oil."

He also believes it's very important that the U.S. own the production of its energy, as well as the management and building of the energy storage systems. "If we create an industry for batteries, solar, wind and nuclear, and we buy all of that from manufacturers outside the U.S., all we're really doing is trading one problem for another," Vieau said. Although such equipment is not something purchased on a daily basis, like oil, it nonetheless represents a substantial industry.

Both Vieau and Riley see China as a dark horse, noting that that nation sees EVs as an opportunity to leapfrog the industry and avoid the 100-year learning curve of how to design top-notch motors and transmissions. "They are very focused on the EV market, very serious," Riley said. "So I'd say it's going to be a race from a mega level between the U.S. and China."

One advantage China has, he said, is that the realities of its political system mean its government is better able to direct economic development in any realm it chooses—in this case, the electrification of transportation, which has been deemed a national priority. Another advantage, Vieau added, is that because so many battery companies are already based there, Chinese companies—BYD in particular, which produces both batteries and plug-in vehicles—have the advantage of in-house demand. "They're not going to have to build their battery businesses by going door to door trying to convince someone to buy their product," he said. "The fact that we don't make much in the way of consumer electronics and PCs in the U.S.—and haven't for some time—is what really stimulated the Asian-based market for batteries and, ultimately, the manufacturing of them."

Nonetheless, both men believe the U.S. stands a good chance of winning the EV race. "We have a tremendous track record on innovation," Vieau said. "A lot of the innovations that spurred the development of lithium-ion batteries in Japan in the early '90s came from U.S. national labs." Riley agreed, adding that the EV industry is America's to lose. "Once America focuses on something and exercises its collective muscles, we're very hard to beat," he said. "We have a unique ability to solve the problems that face us. We have the advantage—when we're focused—of being creative, resourceful, and aggressive, and of having the work ethic necessary to rise to the task. We've done it again and again—the Manhattan Project during World War II, getting a man to the Moon in 1969.

"The challenge now is whether or not we can muster the political and societal will to make this happen."

CHAPTER 14

PLUGGED IN

RICHARD LOWENTHAL, CHARGEPOINT NETWORK

The former California city mayor says he learned long ago that government mandates don't work, and that for any company to be successful, capitalism must have free reign. Now the head of one of the nation's principal producers of EV charging stations and infrastructure, Richard Lowenthal discusses how the goal of American energy independence has driven his interest in EV technologies, the importance of tackling range anxiety, and why U.S. abandonment of manufacturing is a "disease."

As mayor of Cupertino, Calif., for two terms between 2001 and 2006, Richard Lowenthal enjoyed many perks of office. But perhaps his favorite was his government-issued Toyota RAV4 EV, an all-electric version of the company's popular gas-powered SUV. "I absolutely loved the car," he said. "It was quiet and fast and everything was free. The car was free, the fuel was free, the charging station was free."

Produced from 1997 To 2003, the RAV4 EV was Toyota's response to a mandate put out by the California Air Resources Board (CARB) requiring that 2 percent of all cars sold in the state be emission-free by 1998, and 10 percent by 2003. In 2002, however, CARB rescinded the law, and the RAV4 EV was discontinued soon after. Also discontinued was the EV1, GM's answer to the CARB mandate, a car that engendered as passionate a following among its drivers as did the RAV4. Since most RAV4 EVs and all EV1s were available on a lease-only basis, both Toyota and GM were able to collect and crush the vehicles—an act of destruction made infamous by the movie *Who Killed the Electric Car?* Lowenthal's vehicle was one of those towed away and demolished. "When the mandate went away," he said, "the car went away."

The experience left a big impression on Lowenthal, today the CEO and co-founder of Coulomb Technologies, a leading producer of EV charging stations and infrastructure known as the ChargePoint Network. "The market was destroyed when the government changed policies," he said. "My RAV4 was created by government policy and then instantly destroyed by the change of one vote on the air resource board." His conclusion: infrastructure is built by capitalism, not regulation.

Lowenthal served the public for 10 years, twice as mayor, before being termed out of office. The serial entrepreneur and former Cisco executive found himself drawn to the renewable energies arena, and to electric vehicles in particular. He toured Tesla Motors in 2007 and fell in love with the yet-to-be-released Roadster, ordering one that very day. He was particularly struck by the excitement of the engineers, an enthusiasm he found infectious. "They were changing the world and creating a cool product at the same time," Lowenthal said. "I wanted to make a contribution too."

Thinking back to his own plug-in experience, he determined to tackle the elephant in the EV room: range anxiety. "No one wants to talk about it, but it's a huge issue," he said. "I never got stuck with my RAV4 EV, but I had to carefully plan every day. I always had to think." Lowenthal faces the same issue today with his BMW Mini-E, a 100 percent battery-powered EV. "Although I love my Mini, I don't drive from here to San Francisco," which is 50 miles from his Silicon Valley home and office. "The car has a 90-mile range, and unless I can charge there for free, I can't use it. So I decided I wanted to solve that problem—to enable the EV world."

Given his first-hand knowledge of the here-today-gone-tomorrow nature of government policy, Lowenthal was determined to use capitalism to ensure a solid infrastructure for his new business. "We're all networking guys," Lowenthal said of his partner, co-founder Praveen K. Mandal, and the Coulomb team. "The job that needs to be done to make this into an industry is networking. We're proud of our station, we're proud that we're now in the Smithsonian exhibition. But our real business is about doing software systems so that this thing has capitalistic wheels. People have to make money."

The company believes that the success of the plug-in market hinges on supporting an infrastructure that provides overnight charging for those who lack traditional garages, along with public charging locations to enable convenient fueling away from home. Depending on the region of the country, Lowenthal said, only about half of all cars are parked in a garage. "The automakers don't want to talk about the other 50 percent," he said. "But we have to talk about it. We don't want the customer to hesitate on buying an EV because of range anxiety."

While Ray Lane favors GM's extended-range hybrid Volt as the car of the future, Lowenthal believes the all-electric Tesla is on the right track, thanks to a battery big enough to take the car more than 200 miles before needing to be recharged. "Tesla's vision is that the ultimate EV goes 500 miles on a charge," he said. "And that's farther than the human body wants to go. So you'll charge when you sleep and when you work and you're done. How long it takes doesn't matter." He added that the problem with trying to speed up charging time with Level III chargers is that the stations are very expensive. "It's better to have a larger battery and a slow charging time," he said. "We don't need fast-charge, we don't need battery swapping, and we don't need plug-in hybrids." He sees the Volt as a stopgap. "I think it's great and it'll have very high consumer acceptance. But it's really a compromise until batteries are affordable enough that they can switch to the all-electric Tesla model."

And although big batteries currently mean big bucks, Lowenthal predicts that advances in technology will bring the cost of storage down enough by 2020 to make plug-in cars the better choice economically. Gas already costs more than electricity, and it's likely petroleum will only get more expensive. Assuming it does, he said, EVs could make up as much as a quarter of the domestic market by the end of the decade. Lowenthal added that Europe and China will likely embrace electric before Americans do, helped by the fact that in China, at least, there's less of an auto history. "The U.S. will lag a bit," he said. "There's always resistance to change—people are nostalgic and they like to hear the roar of the engine—but we'll get there."

Lowenthal's drive to enable an EV world stems from a desire to see the U.S. independent of foreign oil. Yes, the fact that electricity is cheaper than gas is a plus. Clean air is a plus. "But for me it's all about avoiding imported oil," he said. "The problem with the Iraq War is that we're paying for all the bullets they shoot at us through buying oil. And we're paying for our own bullets. It's ludicrous." Further, he said, the economic effects of oil dependency also need to be to be considered, including the $500 billion trade imbalance in 2009 created by importing petroleum. "I figured the best way I could use my talent was to address the issue of the acceptance of electric vehicles," he said. "If we're successful it sets us free. It sets us free from oil."

One thing Lowenthal doesn't want to see is the U.S. give up on the manufacturing of the EV ecosystem. "It's a disease," he said of the widespread belief that the U.S. can't compete on a production level. "We think of manufacturing as made up of low-wage, uninteresting jobs, but it doesn't have to be that way. We can manufacture with technology." He added that although he's confident the U.S. can stay ahead in the charging industry, his big concern is battery manufacturing. "Do we give up importing oil from the Middle East for importing batteries from China?" he asked. "There's a lot of power in that flow of money." He also emphasized that the U.S. should turn its attention to pursuing other applications for batteries, which he described as "the next big technology."

Not surprisingly, Lowenthal has mixed feelings about the use of government subsidies and credits to stimulate EV adoption. "It accelerates my business so I'll take it," he said. "But I don't believe in subsidies. I like competitive forces." He paused a moment, then continued: "It's complex. If they're trying to create jobs in the U.S., well, that's good. If the goal is to get the volume of EVs—the volume of batteries—up enough so that the price can naturally come down, I'm fine with [subsidies], since high volume is what ultimately will lead to the demise of the gasoline car. Just so long as there's some way this gets relinquished back to the capitalist system."

Lowenthal said what he'd really like to see is a $1 surcharge on the price of gasoline, with the money subsidizing the move to electric. "It's shifting money from one thing that we shouldn't even be doing to something we should," he said. "It's been effective with cigarettes—raise the taxes on cigarettes to pay for anti-smoking campaigns." Lowenthal laughed. "Of course Washington doesn't have the courage to put a fee on gasoline. But that would be the fastest way to make EVs happen. Put a dollar tax on gas and this thing would happen overnight. We saw it in the summer of '08 when gas was $4 a gallon. You couldn't find a Prius and you couldn't sell an Escalade."

Either way, electric is here to stay. "What's different about EVs now is that people want to buy them," Lowenthal said. "The government's not forcing them like it did in the late '90s. And that's what you have to find. You really can't fight capitalism."

CHAPTER 15

Charge America

David Crane, NRG Energy, Inc.

The head of one of the nation's largest power companies talks about why he thinks EVs could make up well over a third of all new car purchases within five years, why plug-ins are the best thing to happen to the energy business since the air conditioner, and why no one needs to be able to drive an EV between New York and LA for the technology to be considered successful.

David Crane is unabashedly optimistic when it comes to predicting EV adoption rates over the next few years. "I think 30 to 40 percent of all new passenger vehicles sold in this country could be plug-in by 2015," said the Houston-based president and CEO of NRG Energy, Inc., a wholesale power generation company.

"We're big believers at this company in destructive technology and innovation," he said. "And we believe the electric vehicle is a classic destructive technology." NRG, whose corporate offices are in Princeton, N.J., is increasingly focused on the alternative energy sector, and has several divisions, including solar and nuclear energy, in addition to Reliant Energy, its retail company.

"When I hear people predict a 2 percent market penetration rate by 2020, I think, 'You have no idea what you're talking about,'" Crane said.

The most aggressive prediction he usually hears is 10 percent. "Just five times more?" He laughed. "That isn't aggressive."

Crane added that he didn't think anyone could truly judge the market penetration for EVs until the electric car is actually here for sale. Had New York City done linear progression modeling in 1880, he said, "They would have estimated the streets would have been six feet deep in horse crap by 1930—they didn't yet have the disruptive technology of the car." Look at how quickly Apple came to dominate the MP3 model when it introduced the iPod, he continued. "I think they achieved a 33 percent market share in something like three or four years." Crane added that recent visits to Nissan and Toyota in Japan left him with the impression those automakers aren't even designing for the internal combustion engine anymore.

According to Crane, the key to getting EVs fully off the ground is to aggressively push the charging infrastructure—though not for the reason most people might think. He said that someone's "got to bite the bullet" and install Level II and Level III charging networks in specific, EV-friendly metropolitan driving regions such as Houston, which has many single-family dwellings with garages. In fact, biting that bullet is exactly what NRG is doing throughout the Houston area, where it plans—on its own dime—to install 50 or 60 strategically located Level III charging stations, which can top up a nearly depleted battery to 80 percent in less than 20 minutes. The company will also install lower-cost Level II charging stations throughout densely populated areas of the metropolitan region, including employee and public parking lots, and is experimenting with a subscription business model similar to those used by the cable and cell phone industries to support both types of stations once they are in place.

What's particularly noteworthy, however, is the way in which Crane views those quick-charge Level III charging stations, which most people consider the equivalent of gas stations. "You can't think about Level III chargers as service stations," he said, noting that most driving needs will be met by nighttime home charging. "That's not what they're there for. Level III chargers are there as an insurance product."

Crane estimated that it will cost NRG—Nissan's launch partner for the LEAF in the Houston metropolitan area—about $3 million to install the Level III charging stations. "That will say to the first wave of electric car buyers, you can use this electric vehicle anywhere in Harris County," he said. "You can blast your air conditioner in the car on a 110-degree day, and you don't have to worry that you're going to run out of charge because we've got a fast charger over in the Galleria or wherever you're headed. Once you assure the early adopters that they can actually use the product without worry, then with all the other incentives that the government is providing, they'll go out and buy the car."

Crane believes the next stage of infrastructure development will move pretty quickly after that, and will come from retailers and restaurant owners. "Anyone who wants 15 to 30 minutes of a high-net-worth individual's captive time—once those retailers realize that there are a lot of people with money in their pockets driving EVs, then people in commercial establishments are going to run out and say to people like us, 'Install Level II chargers in my parking lot, I'm going to offer it as a complimentary benefit.' You don't even have to be an upscale restaurant. You can be a mid-tier restaurant and say to someone, 'Come into my restaurant for an hour and have a $30 meal for two and you can go out and your car will be fully charged.' It'll cost the restaurant maybe 80 cents."

The final stage involves turning EVs into multi-city cars, rather than just commuter and around-town vehicles, he said. NRG has drawn up a map that breaks the U.S. into six urban clusters, with each region encompassing major cities within 200 miles of one another. (The only major cities to fall outside the clusters are Denver and Atlanta.) "With just a little infrastructure among these major cities, you could say, now it's not just an intra-urban vehicle, it's an inter-urban, regional vehicle," Crane said, adding that 180 million of America's 300 million people live within those city clusters. "Then we might need another wave of publicly funded infrastructure development to get people between those cities."

Crane doesn't buy into the idea that EVs need to be able to drive from New York to Los Angeles before they're deemed successful. "That's nonsense, because at the end of the day there are something like 60 million car owners who own multiple vehicles, and I don't think any high-net-worth individual is going to throw out every single one of those internal combustion engine cars," he said. "EV adoption will be more like the way fixed-line phones shifted to mobile phones over the course of 20 or 30 years."

He used his own family to illustrate his point. "I've got five kids and five cars," he said. "I only need one of those five cars to drive from New York to Los Angeles. I like having that capability even though I haven't driven from New York to L.A. in over 20 years."

So is the existing electrical grid up to the challenge? "Absolutely!" Crane said. "We can stay ahead of anything the electric-car world can throw at us. Far from causing the system to collapse, in fact, the electric car will do the opposite—it'll stabilize the system. "The holy grail of the electric industry throughout my entire 20 years in this business has been how to store electricity," he said. "It's the only commodity that can't be stored. Every time someone tries to come up with a super large battery, the damn thing is so damn expensive to build that it makes no economic sense."

Until now. "The genius here—the disruptive technology here—is that if you build the battery to serve a primary purpose, and it can also act as a storage device as its secondary purpose, then doesn't that make sense economically?" he asked. "What we have is a potential for a distributed power storage system."

Even more, he said, it's the ultimate motivation for the development of smart meters. Although unsure whether the vehicle-to-grid platform—in which EV owners sell electricity stored in their cars back to the energy companies during periods of high demand—will happen in his lifetime, Crane said a smart meter system using time-of-use pricing will ultimately prove a boon to the national electrical network.

Asked what drives his work, Crane pointed to several factors. First and foremost, he said, the EV represents the most promising business opportunity for the energy sector since the air conditioner. At the same time, its promise extends far beyond energy. "One of the messages that Washington, D.C., needs to get is that the electric car is a job multiplier," he said. "It's a segment-of-the-economy platform. You start an electric car and a whole world of people will explode who are offering electric car services."

Second, he said, "I'm a proud American, and I don't like sending hundreds of millions of dollars a year overseas for foreign oil." Looking back on his days as a student at the Woodrow Wilson School of Public & International Affairs at Princeton University, where he studied the strategic consequences of the oil crises of 1973 and 1979, Crane said he couldn't believe America was still dealing with the same issue. "If you told me I'd be sitting in the United States 30 years later and we had done absolutely nothing to change dependence on foreign oil, I would have said to you, 'I don't live in that type of country.'"

Now, however, technology is offering a way out. "The battery has come far enough with the Nissan LEAF and its 100-mile range that we now have a way of weaning ourselves off that foreign addiction," he said. "I fundamentally buy the national security argument and public policy reasons why the government should encourage electric vehicles."

Crane also said he takes the climate change argument very seriously. "I'm the father of five children and I don't second guess the scientists of the world when they say we're playing with catastrophic environmental consequences if we don't reduce our carbon by an enormous amount by the year 2050. And I don't know how you get there without bringing the power and the transportation sectors along."

Last but not least: "Who doesn't want to be involved in something that's going to change the way every American lives their life, and change it for the better?"

CHAPTER 16

The Smarter Grid

Charlie Allcock, Portland General Electric

Oregon's largest energy utility is more than equipped to handle any increased electrical load that EVs are likely to impose, says one of its top executives–an assurance that bodes well for widespread adoption of electric vehicles at the national level. In the following chapter, he and another energy specialist elaborate on why they consider the electrical grid a living organism, how economic incentives—and even a bit of guilt—will ensure off-peak charging, and why the electric vehicle is the first smart appliance.

Nissan chose Oregon as a launch pad for its all-electric LEAF for a good reason. Public opinion polls and market surveys consistently show strong evidence of Oregonians' interest in sustainability, and the state has one of the highest per capita rates of Prius ownership in the country. Portland General Electric (PGE), the state's largest utility, is moving forward on a plan to stop burning coal at Oregon's only operating coal plant by 2020, and a regional mandate requires that 25 percent of all grid delivery come from renewable sources five years after that. In addition, nearly 10 percent of PGE's 800,000 customers have volunteered to pay extra for power that is generated from renewable sources, with more signing on each day.

Oregon's selection as one of five test markets for the Nissan LEAF makes it a partner in the Electric Vehicle Project of the Electric Transportation Engineering Corp. (eTec), which was the recipient of a $99.8-million grant from the U.S. Department of Energy in 2009. The company is partnering with Nissan North America to deploy up to 4,700 LEAFs and 11,210 charging stations to support them in selected cities in Oregon, Arizona, California, Tennessee, and Washington.

By mid-2011, Oregon will have at least 2,000 new charging stations throughout Portland, Eugene, and Corvallis to add to the state's original stock of 25. Approximately 900 of the new stations will go to LEAF owners, with the remaining ones dispersed throughout the public domain. The EV Project will collect and analyze data from LEAF drivers to evaluate the impact electric vehicles have on both communities and their electrical infrastructure.

Once the project was announced, PGE immediately began getting calls from customers about installing charging stations, according to Rick Durst, energy information services program manager for the utility. People even filled out public survey forms to request charging spots for their favorite shopping mall. But neither he nor Charlie Allcock, PGE's director of economic development, is worried about the utility's ability to meet a surge in demand. In addition to the fact that there is plenty of excess grid capacity at night, Allcock said, utilities are accustomed to day-to-day and seasonal fluctuations in consumer demand.

Although the public generally sees the electrical grid as a static load curve on a graph, Allcock views it a dynamic system. "The grid is a living animal—an organism," he said. Usage loads rise and fall as businesses expand, shut down, and relocate, and customers install solar panels or initiate other energy conservation programs. PGE is in the process of moving away from coal and natural gas to bring more renewable energy online, particularly wind, which will help keep prices stable even in the face of increased demand.

Nonetheless, Allcock sees the grid as an organism that can be managed. "If people are really going to depend on electricity for fuel, the utility has another set of obligations to deliver, and we've got to perform," Allcock said. "We take that very seriously. But it's not like everyone's going to go and switch to a plug-in tomorrow. We'll see it coming." He noted that PGE experienced a significant ramp-up in electricity use over a two-decade period, from 1989 to 2008, when the percentage of households using air conditioners throughout the utility's service territory jumped from 29 percent to 77 percent. Durst agreed, explaining that consumers have been using more and more electricity despite the introduction of energy-efficient appliances and products. When it comes to increased load on the grid due to data services, air conditioners, and plug-in vehicles—these things don't happen overnight, he said. "We can see the load on our system and plan for it." Durst added that since most people with EVs have meters, PGE can easily monitor the impact of charging. "It's a learning process, but it's no different than what we do with anything else."

Even if within three years half of all new cars in Oregon had a plug—which neither Allcock nor Durst anticipate—PGE could handle the load. "We'd have to be very careful, of course," Allcock said, adding that the utility would encourage customers to be judicious about when they plugged their cars in, urging off-peak charging to limit load on the grid during periods of high use. "It's like the recycling ethic. You guilt people into doing it. You train them. You get them to think when they plug in."

He also expects that either the cars or the home charging stations coming on the market will come equipped with enough computer technology to enable programmed charging for the cheapest time at night, with the utility and charging station sorting out the details electronically to ensure the car is ready at a set time. The best part, he said, is that the system will take much of the thinking out of the equation for the customer. "I can pull into my garage and let the technology do its thing because the plug-in vehicle is the first smart appliance." In addition to being mobile, he noted, this particular smart appliance also has the ability to store energy. "We've tried to get refrigerators and air conditioners to act smart, but this is a different beast altogether."

Allcock sees yet-to-be-implemented regulations as the biggest hurdle facing the utility. "We'd love to know what both the federal and state governments see as the utilities' role, and right now we've got no direction at all," he said. He added that he sees federal and state incentives as especially important to encourage EV adoption. In addition to the $7,500 plug-in buyers receive from the federal government to offset the steep cost associated with a lithium-ion battery, the state of Oregon offers an additional $1,500. "I'd always like to see more incentives, at least in the early years," he said. "The here-today-gone-tomorrow approach is the worst thing you can do to a market. Stability and predictability of rules, regulations, and incentives—that's really important."

Allcock and Durst are big fans of electric vehicles—Durst has been driving an EV for 10 years and pays extra to power his home and car with renewable electricity—and both pointed to the impact of excess CO2 on the planet to explain their enthusiasm for working to support a system that replaces molecules with electrons as a means of fueling transportation. "It's a sense of making a difference and maybe leaving the planet better than when I first found it," Allcock said, noting that the electrification of personal transportation could make a big dent in emissions. "Call it legacy—whatever you want to call it—but that's a big one for me. I want to provide for my family and enjoy a nice lifestyle, but in the end it has to contribute to a greater good."

Like Ray Lane and Richard Lowenthal, Allcock is also intrigued by the possibilities inherent in capturing renewable energy and storing it in a battery for use later, and not just in a car. "The idea of being able to store energy, to store it in different ways, and deliver it to a human creates a much more dynamic environment today," he said, noting that batteries of all kinds could be used to stockpile solar, wind, or any other source of generated electricity. Power generated by a windmill, for example, doesn't need to end when the wind dies. Capture that energy with a battery and renewable power moves into a whole new realm that will extend well beyond plug-ins. "It's on us, whether we recognize it or not," he said, referring to the clean tech revolution. "It's here, and the electric car is going to accelerate it."

CHAPTER 17

NATIONAL ENERGY

SECRETARY STEVEN CHU, U.S. DEPARTMENT OF ENERGY

Energy Secretary Steven Chu has been outspoken about the need for the federal government to invest heavily in EVs and other alternative energy industries as a way to help America achieve energy independence, boost the economy, and lessen the effects of climate change. He has also said he would like to see the nation's top innovators working together to tackle the energy crisis, just as a previous generation of scientists came together on the Manhattan Project to counter an earlier global threat.

Steven Chu may have entered the world of politics with his appointment to President Obama's Cabinet, but the energy secretary is first and foremost a scientist. The co-winner of the Nobel Prize in Physics in 1997, Chu has long been consumed with technologies to help combat climate change. Before being tapped to head the Department of Energy, Chu was the director of Lawrence Berkeley National Laboratory in California, where he led a team of scientists pursuing breakthroughs in renewable energy sources, with the twin goals of reducing greenhouse gas emissions and weaning the world off fossil fuels.

Since taking over the Department of Energy, however, Chu has expanded his focus to include the economic and national security aspects of the growing energy crisis. He has repeatedly called for the U.S. to step up its investment in renewable sources of energy or risk falling permanently behind other nations in the race to develop electric vehicles and other alternative energy technologies. In a speech at the Colorado Energy Jobs Summit at the University of Colorado in early 2010, for example, he claimed that China is investing more than $100 billion a year in clean technologies—compared to America's 2009 investment of $18.6 billion, according to the Pew Environment Group—and warned that failure to move quickly could see the U.S. exchanging a dependence on foreign oil for a dependence on foreign technology.

Chu reiterated that message the following month at Stanford University, telling a full-capacity crowd that the federal government should invest "tens of billions of dollars" a year to develop America's nascent clean energy economy. "The overwhelming scientific consensus is that humans are altering the destiny of the planet," he said, adding that while it's not too late to mitigate the worst effects of excess CO2, we can't afford to wait. "If we plow on as usual it could be catastrophic," he said. "It could be very bad. Or very, very bad."

Continuing with the same fossil-fuel-burning energy infrastructure is simply a way of "holding off the inevitable," Chu told the 1,700 students and faculty gathered to hear the former Stanford professor. But holding off the inevitable for another 5 or even 10 years is a lose-lose situation. Not just from an environmental standpoint, he said, but also because it gives other countries a chance to pull ahead in clean tech, forcing the U.S. to play catch-up and putting the nation at risk. "Energy touches everything in the United States," he continued. "It's a new industrial revolution and it's essential for American competitiveness."

Chu used his Stanford address to call for a federally backed, Manhattan-Project-like undertaking to develop a clean energy revolution. "If you look at the amount of funding for [the Manhattan Project], and the amount of funding to put a man on the Moon, it was a huge spike in funding," he said. "I think we need to do that." While the economic stimulus package enacted in February 2009 was a good start, Chu said it wasn't close to enough. "You still need, I think, tens of billions of dollars as a minimum per year invested in these technologies and the associated science."

Scientific innovation is the only way to confront the huge scale of the energy challenge, Chu believes, and the importance of gathering America's top scientists together in a federal program dedicated to tackling greenhouse emissions is something he has spoken of in the past. In a 2009 interview with *National Geographic*, he emphasized the necessity of a dedicated and coordinated scientific effort. "Just like during World War II, when a lot of the best physicists went to work on radar and the atomic bomb, the world needs scientists to work on this issue," he told the magazine. "We're in a war to save our planet."

Although the Obama administration is pursuing multiple avenues in terms of alternative transportation fuels, Chu is known to be a strong electric vehicle advocate. In a 2009 commencement speech at the California Institute of Technology, Chu urged graduates to prepare for the "inevitable transition to electricity as the energy for our personal transportation." A few months later, speaking of the billions of stimulus dollars being distributed by his department in support of clean-energy transportation, including the development of advanced EV batteries, he reportedly told a closed-door alternative energy fuels industry gathering: "If it were up to me, I would put every cent into electric cars."

CHAPTER 18

Houston, We Have a Solution

Mayor Annise Parker

When it comes to public policy, attention is most often paid to the top echelons of government, both federal and state. In the end, however, the nuts and bolts of implementation often takes place at the local level—progress pushing its way up rather than filtering down. Houston's new mayor has just that sort of progress in mind. Although mindful of her city's oil legacy, she is working hard to guide her constituents onto a more energy-efficient path—one that includes a heavy emphasis on electric vehicles.

Any list of America's greenest cities generally includes plenty of familiar names, metropolitan areas already well known for their eco credentials: Portland, Ore., San Francisco, Seattle, Denver, and Austin, Tex., among many others. If Mayor Annise Parker of Houston has her way, however, those municipalities will have to make way for a new addition to their ranks.

Houston?

Yes, Houston. Although America's fourth largest city—and Texas' biggest—is far better known for oil than it is for clean tech, Houston's new mayor is committed to changing that image. At the same time, however, she isn't waiting around for the federal government to take the lead. Like a lot of officials throughout the United States, Parker knows that if cities like hers are dedicated to pursuing an alternative energy agenda, success requires initiative on the part of local governments. She also knows that such initiative has to have the backing of the people.

It's not just a matter of responding to what people want, Parker said, it's also about leading them in the right direction. "You can't force them. You can't bludgeon them. And you can't lead the market too far. I want the city of Houston to be known as a city that recognizes that our bread and butter in the past was the oil and gas industry. But because of the mindset of Houstonians, we may be able to make great advances sooner and sustain them longer than a lot of other cities. We may be considered a very unlikely place for this to happen, but it's this combination of vision and practicality that has taken us forward many times in the past."

Anybody who's spent any time in Houston knows that it's a city that looks to the future, said Parker, who made *Time* magazine's 2010 list of "the people who most affect our world." She appears in the magazine's "Leaders" section, where she ranks 13th on the list. "We're a city that's all about change and we're always interested in new ideas. We'll try them. And if they don't work we'll discard them and go on to the next one."

Although quick to credit the oil and gas industry for fueling Houston's growth throughout much of its history, and for funding the research and development of alternative energy sources in recent years, Parker has not been shy about the importance of clean tech as part of Houston's future. "Oil and gas are going to be with us for a very long time," she said. "But we can see that our needs are changing and the needs of the world are changing, so let's not wait until we've exhausted all the possibilities on the oil and gas side before we explore the possibility of the green and renewable side."

To that end, she said, Houston has been a major purchaser of wind energy, and is working to become a front-runner in Leadership in Energy and Environmental Design (LEED), an internationally recognized green building certification system. "Houston is considered the most air-conditioned city in the world," Parker said. "So if you're out to make a statement—to make a difference in how you design your buildings for energy efficiency, let's do it in Houston! Let's not do it where it's easy. Let's do it where it's a challenge. But in that challenge there's also an opportunity. Because Houston is growing so rapidly and constantly reinventing itself, it's easy to bring in something new and change the landscape."

Despite Houston's strong oil heritage—Parker herself worked in the oil and gas industry for 20 years, mostly at Mosbacher Energy, where she was a software analyst—the mayor said the city has been interested in alternative energy for some time. She noted that Houston purchased a large number of Priuses as government fleet vehicles when Toyota first introduced its groundbreaking hybrid in the late 1990s. Now, with her prompting, Houston has turned its attention to the all-electric Nissan LEAF, which Parker would like to see power the city's next-generation municipal fleet. "That'll give us a chance to put the car through its paces," she said. "Not only will it help our air quality, but we can show the public what it can do. We can help build a market for the LEAF."

Parker said that Houston is a particularly good city to showcase the benefits of the LEAF and other plug-ins. As NRG's David Crane pointed out, not only is Houston a city with many single-family homes—most equipped with garages, which enables easy nighttime charging—it is also the center of a large and widespread metropolitan region that is almost completely reliant on the automobile for its transportation needs. Although the city has been investing in rail and other forms of public transit, Houston is difficult to navigate without a car.

"We are a horizontal city," Parker said. "If you want to get from one side of Houston to the other, you need a personal vehicle. That's why I'm excited about electric vehicles as personal vehicles. Here, a personal vehicle is not a status symbol; it's a necessity." And because it's a necessity, she said, Houston needs to think about what impact that personal vehicle has on the environment.

When asked if she thought gasoline should be taxed at a higher rate, Parker paused, carefully choosing her words before answering that she did. "I think that we need to make a clear link in the public's mind that the roads we drive on are not free," she said. "There are services that are linked to the consumption of fuel, which is a nonrenewable resource."

Although Parker is a strong environmentalist—she describes herself as an avid recycler and a long-time member of the Sierra Club and the Nature Conservancy—she is also a realist. For all the good environmental reasons to support electric vehicles, including achieving challenging air quality goals, she said, any change has to fit with existing lifestyles.

"It needs to be convenient, it needs to be practical, and it needs to strike that cord in us that says we can do this, and we can do this the Houston way," she said of the move to EV technology. "We all know that electric vehicles and other forms of alternative transportation are the future. But we can't tell people that. We have to show them."

CHAPTER 19

CHINA'S CHARGE

WANG CHUAN-FU, BYD AUTO

China's up-and-coming version of Bill Gates has always been ahead of the curve, and American entrepreneurs can learn a thing or two from China's wealthiest citizen and leading wunderkind. The head of one of its preeminent electric car companies doesn't just put his money where his mouth is. He also put his mouth where his money is, downing a glass of battery fluid to show how nontoxic lithium-ion really is.

T*ime* magazine didn't exactly go out on a limb when it named Microsoft co-founder Bill Gates one of the most influential people of the twentieth century. The pick was pretty much a no-brainer. Not only did Gates see the potential of the personal computer revolution long before most, he also had the technical genius and entrepreneurial chops to do something about it. Microsoft's success launched a technology and economic boon the likes of which the world had never before seen, ultimately making Gates the wealthiest man in the world for many years running.

Wang Chuan-Fu, CEO of China-based BYD Auto, could very well lead any comparable twenty-first century list. BYD, which stands for Build Your Dreams, recently topped *Businessweek* magazine's Tech 100 list, where it outranked Apple and Amazon. Like Gates in late-twentieth-century America, Wang is poised to bring incredible wealth not only to himself, but also to the Chinese economy.

"This guy is a combination of Thomas Edison and Jack Welch," said Charlie Munger, the 85-year-old vice chairman of Berkshire Hathaway, in an interview with *Fortune* magazine, in which he compared the workaholic Wang to the famous American inventor and to the former chairman and CEO of General Electric. "Something like Edison in solving technical problems, and something like Welch in getting done what he needs to do. I have never seen anything like it."

Munger isn't an easy man to impress. So when Munger sang Wang's praises, Warren Buffet, the legendary head of Berkshire Hathaway, listened to his longtime partner and good friend. Although Buffet is well known for avoiding investments in businesses he doesn't already know and follow—he famously gave wide berth to the technology industry of the 1990s—he was intrigued enough to take a look. After careful research, Buffet decided he liked what he saw, and in late 2008 a subsidiary of Berkshire Hathaway paid $230 million for a 10 percent interest in BYD.

There is no question that Buffet has been pleased with his investment. In fact, the Berkshire Hathaway chief has said his only mistake was buying a 10 percent stake in BYD when he should have bought 20 percent.

Why? Because in 20 years, all cars on the road will be electric, Buffet told the *Houston Chronicle* in November 2009.

Berkshire Hathaway's investment catapulted BYD into the big leagues, taking Wang along with it. Thanks to Buffet's stake, which led to a fivefold leap in BYD's shares, Wang became the richest person in China in 2009, with a net worth of $5.1 billion. By all accounts, however, increased wealth hasn't affected Wang's habits much. He still eats in the company cafeteria and lives in a company-owned housing complex not far from his office.

An engineer by training, Wang got his entrepreneurial start studying the patents of other companies' mobile phone batteries, even going so far as to take them apart to see how they were made. In 1995, at the age of 29, the former government researcher started his own mobile phone battery-making business in Shenzhen, the special economic zone just north of Hong Kong. Today BYD is one of the world's largest battery producers, supplying Nokia, Samsung, and Motorola.

Wang, who was born to a family of poor farmers, and whose parents both died when he was a teenager, entered the automobile business in 2003. Although he knew little about making cars, Wang bought a failing Chinese state-owned car company. Within six years BYD's model had become China's bestselling sedan. Today BYD offers a number of different cars in China, including a plug-hybrid vehicle that at $24,900 is cheaper than the Toyota Prius or the soon-to-be-released GM-Volt. In addition, Daimler AG recently announced a joint venture with BYD to develop all-electric cars for China.

Wang, however, has no intention of stopping at China's borders. In fact, he intends to make BYD, which currently employs 130,000 workers, the world's biggest carmaker by 2025. "It is a big ambition," Wang told CNN in a 2009 interview, though few, if any, who have watched his meteoric rise, believe he is intimidated by the challenge. Within the next year, BYD plans to sell EVs in the U.S.—the company is opening its North American headquarters in Los Angeles—and both hybrids and EVs in Europe.

And while BYD's current cars, like many Chinese-made products, may not yet be up to Western standards in terms of quality, luxury, and performance, Wang watchers like Buffet don't expect that to remain the case for long. In fact, Wang is determined to create a high-quality EV that will grab the attention and respect of the world. "For new-energy vehicles, China is on the same level or even leading other countries," he said in his CNN interview. "In the field of new-energy cars, China hopes that Chinese companies can catch up with the rest of the world."

Of particular note here is that Wang refers to what China wants. In fact, when it comes to EVs, Wang and the Chinese government want the same thing: a transportation system based on plug-ins—and not just because they offer a badly needed solution to China's pollution crisis. Of equal, perhaps even greater, importance are the phenomenal economic benefits that will grow from the resulting electriconomy—benefits that can be reaped both domestically and abroad.

That shared goal gives Wang a significant leg up on the global competition, leaving BYD—and China as a whole—in a position to leapfrog the U.S. and other nations when it comes to alternative energy industries. (BYD is perfectly positioned to benefit from China's electric car subsidies and government mandates regarding charging stations.) As stated earlier, China is currently pouring more than 10 times the resources into EVs and other renewable energy industries compared to the U.S. Equally important, however, is China's sense of urgency around its goal. While the Obama administration is calling for one million plug-in vehicles by 2015, China has mandated that the same number of EV and fuel-efficient vehicles be on Chinese roads by 2012. Thanks to government grants and initiatives, as well as large and targeted investments, China is well on its way to exceeding its goal.

Wang shares yet another trait with Gates, one that enables him to look beyond wealth and business success. In January 2000, Gates made the decision to step down as CEO of Microsoft (he remains company chairman) to focus his attention on the Bill & Melinda Gates Foundation, his philanthropic organization. And although Wang is just getting started on his global aspirations and is likely to continue as a strong presence in BYD for many years to come, he, too, seems increasingly concerned with society as a whole. Wang concluded his CNN interview by pointing to the overwhelming environmental issues China faces as the world's biggest producer of greenhouse gases.

"As an entrepreneur, I think I have to consider both aspects," he said. "One part is the creation of a new business mode, or the revelation of new business competition. The other is that it's for social responsibility, making our earth bluer. Urban pollution, reliance on petroleum, and emission of carbon dioxide are three problems that entrepreneurs have to consider for basic social responsibility."

PART III

The Road Forward

"While I take inspiration from the past, like most Americans, I live for the future."

—Ronald Reagan

"When it comes to the future, there are three kinds of people: those who let it happen, those who make it happen, and those who wonder what happened."

—Richard M. Richardson Jr.,
American University professor

CHAPTER 20

FREE GAS

Roger Brent's Northern California home offers a window into the future. It's also the perfect answer to anyone who erroneously claims that electric vehicles simply transfer the tailpipe from the car to the smokestack. Even when dirty coal is used to produce the electricity, EVs are always more efficient, and thus cleaner, than gas-powered cars. In Roger's case, however, there is no smokestack — not for the car, not for the house, and not for the energy used in the house.

Perched high in the Berkeley hills, Brent's home features a postcard-worthy view of the East Bay, San Francisco, and the famous body of water that separates them. It also features 400 square feet of indigo-colored solar paneling atop its roof, which produces enough energy year round to power the two-story house and the Tesla Roadster in the carport. (The Tesla Roadster is an all-electric ultra-high performance sports car with a stated range of up to 245 miles and blow-your-hair-back acceleration.) A member in the Division of Basic Science at the Fred Hutchinson Cancer Research Center in Seattle, an affiliate professor in the Department of Genome Sciences at the University of Washington, and an adjunct professor of Biopharmaceutical Sciences at the University of California in San Francisco, Brent shares the home with his wife, Linda Buck, a Nobel-Prize-winning scientist.

Like any number of other smart, well-off people, Brent and Buck are better able to live in the future. When Microsoft co-founder Bill Gates built his house 15 years ago, his liquid crystal display (LCD) TV represented a scene from the future. Now the same LCD picture gallery can be stuffed in a Costco shopping cart for a few hundred dollars. Similarly, Paul Allen, the other smart—and also very wealthy—Microsoft co-founder, enjoyed the equivalent of a TiVo, the first digital video recorder (DVR), three years before TiVo Inc. opened its doors in 1997. The system probably cost Allen a million dollars. Today, however, anyone can get an equivalent DVR for free by signing up for digital cable or Direct TV.

While certainly no Bill Gates or Paul Allen in terms of wealth, the Berkeley couple's circumstances allow them to be early adopters of technologies others might not be able to enjoy until a few years down the road, when mass production and economies of scale bring lower prices.

But Brent's interest in homemade energy production extends far beyond simply being an early adopter of a new technology. Although driven in part by the national security aspects of energy dependency, he was primarily concerned with finding a way to limit his carbon footprint. A brilliant and successful biologist who regularly travels the world for business and pleasure, Brent notes that he burns more than his weight in jet fuel each month. "My work consumes money and resources like crazy," he said, although he is well aware that it also gives back in terms of scientific research and life-saving drugs.

His solar energy system is yet another way to give back. "I realized my house was a poster-child electricity factory and that it was a waste of a good resources not to do it," he said. "I realized I should put my money where my mouth is, so we're shifting to more renewable energy resources when possible. Some of the motivation here is that it's the right thing to do. A man should try to cover his own footprint."

Or, at the very least, the footprint of his $100,000 sports car, even if it is electric. In 2007, Brent decided to see if the economics of going solar made sense. Scientist that he is, he said he "geeked up" on the technology when determining what to install, eventually opting for a photovoltaic solar-power system. Rooftop solar panels convert sunlight directly into electricity, producing as much as 5,000 watts during daylight hours—though the average is closer to 3,000 watts. Brent says he saves $200 to $300 each month in electricity costs and his bills always show a credit. He added that his setup provides more energy than originally anticipated, even in a rainstorm, and is capable of feeding back in to the grid if necessary.

"I realized having the system in place would make the Tesla operation costs zero," he said. "I went through the 'If not me, who?' and 'If not now, when?' Rabbi Hillel argument," he continued, referring to the first-century sage. "I couldn't really afford it, but what the heck." In addition to what he spent on the Tesla, Brent estimates that after taking advantage of the solar investment tax credit California provides, he paid $28,000 for the system, including the cost of upgrading the home's electric service to 200 amps.

Despite his initial reservations, Brent clearly enjoys his car. It is sleek, black, and head-turning sexy. Behind the wheel he (mostly) discards his scientist hat, relaxes into the seat, and revels in the power and performance of what is arguably the hottest car on the road today. "It's really fast," he said. "She goes where you point her." But as pleasurable as the Tesla is to drive, Brent views it with humility. "Nobody deserves a car this wonderful."

Although he talks about putting his money where his mouth is, Brent emphasized he isn't obsessed with living a sustainable life. He loves living off the grid but he's no ascetic. "I like to live large," he said. "I like my pretty house. I like my car. I like my hot tub. I like cooking and eating great meals. I like going to other countries and learning new things and meeting extraordinary people." For him, however, living large at home entails limited environmental impact.

Brent firmly believes government grants and tax credits are necessary to support the development and adoption of renewable energy technologies, particularly in the midst of a recession. He points to what he sees as top-notch engineering behind his Fronius IG inverter, which converts the electricity from the solar panels to a grid-friendly alternating current, and credited the German government with having the vision to support the design of clean tech in its early days. He also notes that government subsidies were instrumental in launching California's solar industry, and gives both the state and U.S. Energy Secretary Steven Chu high marks for pursuing that route, adding that more will be required to accomplish a large-scale shift to EVs. "Tax credits are huge," he said. "Even in the face of the recession, solar makes good sense."

It also makes good sense in terms of national security and energy independence. The fuel that powers Brent's home and car is not just made in the USA, it is literally made at home, and for no ongoing cost. Over the next 20 years, as more homeowners adopt "personal power plants" and use the power generated to fuel their transportation, we will not only have eliminated the need for imported oil, but we will also have organically created a distributed electric power grid nearly impossible to disable via terrorist attack. To speed that outcome, *JOLT!* suggests (only half in jest) creating another amendment to the United States Constitution, one written along the lines of the Second Amendment: "A well regulated electric utility, being necessary to the security of a free State, the right of the people to harness and produce electricity shall not be infringed."

Looking out at the evening constellation of lights stretching beyond the airy rooms of the zero-energy home he loves, Brent looks forward to the day when the buildings emitting those lights, and the cars in their carports and garages, will be powered with alternative energy sources, just as his house and car are today. In five years, he predicted, significant areas of the country will be well populated with EVs. He anticipated their drivers discovering what he felt when he first drove his all-electric car: "The unexpected feeling of finally living in the twenty-first century and having a twenty-first-century experience."

"Learn the past, watch the present, and create the future."

—Jesse Conrad, author

CHAPTER 21

BACK TO THE FUTURE

Roger Brent's "free gas" setup represents the future, but it's a future that began a long time ago. The "overnight" success of all technology revolutions, including the computer and the Internet, always entails years—even decades—of hard work and innovation, and EVs are no exception.

The first electric vehicles appeared on U.S. roads in the 1890s and quickly powered past the internal combustion engine in popularity. By the end of the decade, EVs outsold gasoline cars 10 to 1. Of the nearly 4,200 cars produced in the United States by 1900, almost 30 percent ran on electricity—with steam and gas powering the others—and electric vehicles made up about one third of all cars in New York City, Boston, and Chicago. Their appeal grew out of the fact that they were clean, quiet, and reliable; electric vehicles, in fact, were the only ones able to operate during a blizzard that hit New York City in 1898. Gas cars, in contrast, were smelly and noisy, and required wrestling with gears and working a balky hand crank to start the engine. In addition, the few roads of the time were located in and around towns and cities. As a result, most trips were limited to local commutes, a situation that favored the shorter range of the electric vehicle, particularly in light of the scarcity of fuel in rural areas.

Ultimately, however, the collapse of the financial markets in 1907, exacerbated by the flood of money sent to San Francisco following its devastating earthquake the previous year, made it impossible for the one company offering a viable battery to bankroll increased production. The shaky economics of the era also halted all funding of the charging infrastructure needed to launch the EV on a larger scale.

Things only got worse for the electric vehicle with the increasingly widespread distribution of plentiful and cheap oil from the Spindletop Hill oil field in Texas, located just outside Beaumont. Years of patient exploration on the part of a Sunday school teacher and land speculator named Pattillo Higgins paid off on the morning of January 10, 1901, when a 200-foot-high eruption of petroleum stained the Texas sky. The geyser signaled the end of John D. Rockefeller's Standard Oil monopoly and inaugurated the oil age. Spindletop quickly became the most productive field the world had ever seen, causing oil prices in some areas to drop below those of water. (Despite the oil field's initial promise, however, production at Spindletop fell dramatically after two years. A later drill in the area in 1925 led to a second oil boom, which peaked in 1927. By 1936 Spindletop was essentially dry.)

The death knell for widespread adoption of EVs came in the form of an ambitious entrepreneur and automobile builder named Henry Ford, who took one look at the price of oil and determined to bring the internal combustion engine to the masses. The Model T hit the streets in 1908, its simple design and assembly-line production making the car's $500 to $1,000 price tag affordable for thousands of Americans, especially when compared to the cost of the larger and more opulent electric vehicles, which by 1910 ran as much as $3,000.

With gas cheaper than electricity, the internal combustion engine was off and running. It got another boost after 1912 when the awkward hand crank was replaced with an electric starter. The next decade saw the expansion of the nation's roadway system, further pushing buyers toward combustion engines, which could more easily travel the long distances between cities. By the 1920s inexpensive, readily available gas combined with the longer range and greater horsepower of the gas-powered engine had pushed the EV from the national stage.

But the cars retained their fan base—including the wife of Henry Ford, who drove a 1914 Detroit Electric—and never completely disappeared from the scene. Electric enthusiasts kept the EV flame alive, jumping on the few golf-cart lookalikes that were released in the late 1960s and early 1970s, or working at home to convert their conventional cars to electric. And just like Apple Computer founders Steve Jobs and Steve Wozniak in the earliest days of the personal computer, they formed local clubs and met on weekend afternoons to trade tips, tinker on motors, and customize their cars.

It wasn't until the introduction of General Motor's lease-only EV1 in 1997 that electric vehicles moved back into public consciousness, if only at the fringe. The stars seemed to have aligned for the EV1: it was well engineered, the second-generation model featured a superior nickel-metal-hydride (NiMH) battery pack that powered the car up to 75 miles on one charge, and it had the world's biggest car company behind it. Developed in response to California's 1990 zero-emissions vehicle mandate, which was subsequently rescinded in 2002, the lease-only car proved wildly popular with its drivers, even though leases were expensive and proprietary charging stations few and far between.

The EV1 was expensive to produce, however, and the financially strapped company ended the program after a few years, when California revoked its regulation requiring zero-emission vehicles. GM claimed the battery didn't work well in cold weather. It also claimed there wasn't enough demand, although after the EV1 assembly line stopped production in 1999, a waiting list of 5,000 names was made public. In any case, the company defended its decision by stating that it could not support such a limited number of complex and novel vehicles for a reasonable cost. GM recalled all EV1s in 2003 and crushed them two years later, prompting outrage and bitter disappointment among the vehicle's many fans, who had tried in vain to convince the company to let them purchase the cars.

Although other major car companies released a small number of lease-only electric cars around the same time as GM, the only other vehicle that made significant inroads with the American public was Toyota's RAV4 EV, the model Richard Lowenthal drove as mayor of Cupertino. Produced between 1997 and 2003, the RAV4 EV achieved a top speed of 78 miles per hour and had a range of up to 120 miles. As with the EV1, RAV4 EV drivers were passionate about their cars. The vehicle, however, featured one significant drawback: drivers had to buy a wall-mounted charging unit, which made the already expensive vehicle that much more expensive. The several hundred cars that ultimately became available for purchase had a sticker price of over $40,000.

Toyota recalled its leased vehicles in 2003—like GM, the company refused to sell to drivers who wanted them—and crushed many before yielding to a campaign to halt the destruction. Many of the people who managed to buy the vehicles in 2002, when the RAV4 EV was briefly available for sale, have held on to them ever since, and any cars that find their way onto the market are quickly snapped up. And although the RAV4 EV's NiMH batteries weren't expected to last beyond 10 years, many of the original batteries in the early-year models are still going strong.

In addition to a lack of corporate enthusiasm, a problem dogging both the EV1 and the RAV4 EV was the absence of a standard plug. Unable to agree on an industry standard, each EV manufacturer produced its own, making charging the car anywhere other than home challenging. (Consider how difficult it would be to buy a refrigerator if each brand manufactured its own plug.) The good news is that the next-generation EVs will enjoy standardization from the get-go. The Society of Automotive Engineers has developed a five-pin connector that can be used with either 120-volt or 240-volt electrical systems, meaning all charging stations will easily connect to EVs, just as all gas station nozzles fit conventional cars.

Every major automaker that builds cars in North America or exports them here has agreed to abide by the same-plug standard, known as the SAE J1772. Automakers launching plug-in cars in the United States within the next 12 months include Ford, General Motors, Audi, BYD (China), Coda (China), Fisker, Nissan, Toyota, Smart, Rolls Royce, BMW, Tesla, Fiat, and Mitsubishi. In addition to offering simplified charging, each company will enjoy lower manufacturing costs as a result of its decision to support a standard plug.

Though pushed deep underground a hundred years ago by Henry Ford's Model T and the ready availability of cheap oil, the electric car endured, thanks in no small part to its dedicated fan base. Long perceived as a group of oddball fanatics, their day has finally come as the EV moves firmly into the mainstream.

"I think there is a world market for maybe five computers."

—Remark attributed to Thomas J. Watson, Sr., Chairman of the Board of IBM, 1943

CHAPTER 22

ENERGIZE YOUR MOTORS

"For the first time since the turn of the last century, the U.S. is seeing the mass introduction of the electric vehicle. Within the next two years, nearly every major car manufacturer is planning to release either a plug-in hybrid or an all-electric vehicle, with at least five new models scheduled to hit the streets by the end of 2010. No longer prototypes or regulation-appeasing tokens, these are real cars being put out by real, profit-minded companies, with the dealer and parts networks to support them indefinitely.

The typical range of the all-electric vehicles being released over the next couple of years will be around 100 miles, and it will take a bit of time before the public charging infrastructure to support battery-only cars is fully deployed. However, it's important to realize that charging infrastructure is not the barrier to EV adoption that many believe. While there are some 150,000 gas stations in the U.S., there are around 110 million households, each with the capacity to charge a car. And about 65 million of those households have a garage with one or more parking bays.

Generally speaking, the people who buy new cars have disposable income. They also tend to have homes with garages. It is therefore likely that the 60 million households with two or more cars will also be the first to buy new EVs as a second or even third vehicle. Those second or third cars will become the around-town or commuter cars for the household, with 80 percent of the charging done at home.

TWO CARS IN EVERY GARAGE

U.S. Households With Two or More Cars and a Garage

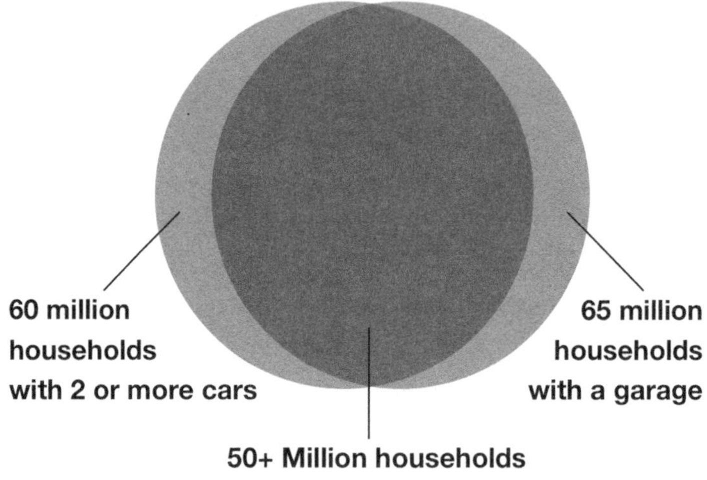

A majority of Americans with two or more cars also have a garage or carport to park them in. Made up of more than 50 million households, this group is a perfect match for the electric vehicles soon to come on the market.

In actuality, this set-up represents another form of "hybrid," in this case divided between two cars—the ICE for longer trips and the EV for the bulk of the household transportation needs, which as we already know, make up of 90 percent of all driving miles. Theoretically speaking, we could replace 60 million gas-powered second cars—25 percent of the U.S. auto population—with EVs, all without installing even one public charging station.

One-car households, on the other hand, will turn to plug-in hybrids like the GM-Volt, which will allow those drivers to cover all their transportation requirements, and still use electric fuel for all but 10 percent of their driving miles. Down the road, however, as the range of all-battery electric vehicles climbs to 200 and even 300 miles, with rapid-charging infrastructure along roads and interstate highways growing to meet demand, sales will begin to shift definitively to all-electric models.

Citizens of the Pacific Northwest will be some of the first to see this future without even having to wait for battery improvements, since public charging build-out is already taking place wherever EV companies are targeting their largest initial deployments. As a result, in addition to the plans Oregon has in place as a launch site for the LEAF, Washington State will be the first to electrify the entire length of its state. By mid-2011, the west side of state, where the vast majority of the population lives, will have 2,000 Level II and Level III public charging stations spread throughout major city centers, as well as along its 300 miles of interstate highway, which will have Level III charging locations set up every 40 miles.

Breakthroughs in lithium-ion battery technology have brought greater energy storage in a smaller, lighter package, thus enabling the EV to cross a crucial threshold of feasibility for use in a car. That threshold represents a turning point with clear parallels to the Internet revolution, which also resulted in a massive paradigm shift. While the 9,000 lines of code Marc Andreessen co-authored to create Mosaic, the first widely used Web browser, led to the creation of the Internet as we know it today, those 9,000 lines didn't invent the Internet. Andreessen's seminal code sat atop millions of lines of code written by many others in the decades prior to his breakthrough, just as current advances in lithium-ion technology grew from decades of earlier developments in energy storage. And since the best predictor of future events is the past, it's worth considering the Internet comparison more closely.

As with the Internet, widespread adoption of the EV will have a profound impact on many, with plenty of winners and losers. The Internet created far more winners than losers, with gains that have been felt globally in terms information, economics, the environment, and society in general—with a particularly strong impact on the consumer. In a repeat of the earliest days of the Internet, the technology and infrastructure needed to launch the EV and its associated industries is already in place. The electric car is now at a tipping point that will ignite a revolution, opening the door to massive investment opportunities that will spawn vast economic growth and see the creation of powerhouse companies. Only a few people anticipated Google, eBay, Amazon, and Facebook; instead, the Internet grew organically, driven by consumer demand. The best predictors of the Internet and its huge impact on the world came from forward-looking venture capital firms and government research agencies focused on where the market was headed and that placed their money accordingly. The same is true for EVs and their ecosystems today.

EVs WILL DOMINATE IN LESS THAN A GENERATION

Pace of Technology Shift

$$\frac{1G}{\left(W/L\right)^2} = \text{Pace of Shift}$$

This "right-brain" equation postulates that the pace of a technology shift within a society can be predicted based upon the ratio of winners (those who benefit) to losers (people and companies that currently benefit from the status quo).

1G = one generation, or about 20 to 25 years
W = winners
L = losers

And although the Internet seemed to come out of nowhere in the mid-1990s, its relatively rapid adoption was the result of years of innovation and hard work. Again, the same is true of EVs, whose batteries, sophisticated electronics, lightweight materials, electric motors, and regenerative braking, as well as the smart-charging infrastructure that supports them, represent countless years of innovation and experimentation.

Why was the Internet so rapidly and widely adopted? Yes, it provides access to information, entertainment, commerce, education, and research—but we had all those things before the Internet ever existed.

The answer? Three simple words:

- Better
- Faster
- Cheaper

These three words sum up all important technology market shifts. The Internet offered significant improvement on all three counts. And so do EVs.

Just like the Internet, which offers countless cost-saving benefits for businesses and other organizations, electric cars are both cheaper to own and operate than gas-powered vehicles—a price gap that will only widen with the passage of time. And like the data that travels across the Internet, the fuel that powers EVs moves along the electrical grid at near the speed of light, which is at least a million times faster than shipping oil across the ocean. In addition, both industries allow for the disintermediation of the market, where the middleman is often cut from the equation. Buying electricity from the utility, for example, is comparable to getting gas directly from Exxon's refinery.

Of course, "better" is subjective. There is much that is better about an EV compared to a gas-powered car. Some will buy an electric car because it's more fun to drive. Others because it's as quiet as a Rolls-Royce. And still others will buy it because it's "greener." As Americans, however, we can all agree that creating a stronger and more secure nation—and not sending jobs and cash to the Middle East—is definitely better than continuing along our current path.

Almost no one saw the Internet coming. And of the very few who did anticipate the technology, nearly all misjudged how quickly it would grow. Throughout the Internet's first years, some of the industry's most respected pundits regularly called for its demise, predicting the infrastructure would essentially collapse under the weight of its own success.

Sound familiar?

EXPERTS WRONG AGAIN

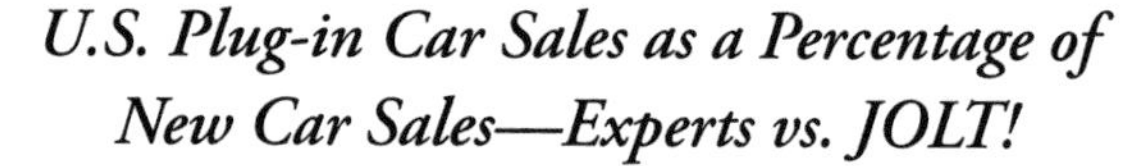

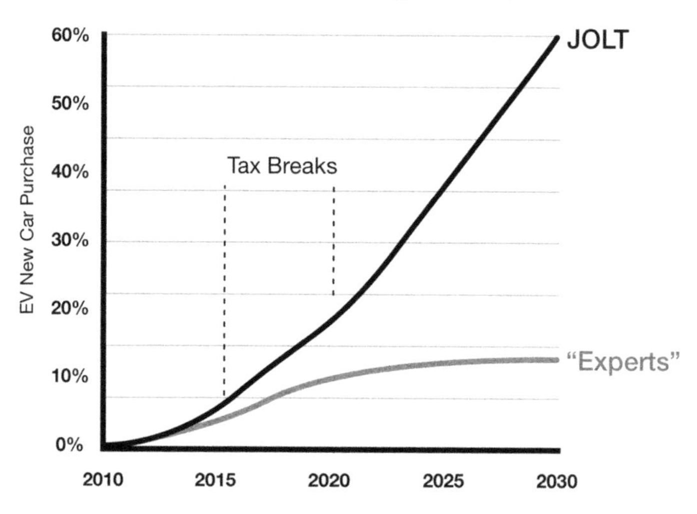

JOLT! predicts a much higher rate of EV adoption than do other "experts." Spurred by government subsidies, electric cars will make up 60 percent of all new car sales by 2030. Despite the substantial and ever-increasing technological advantages of an electric drivetrain, however, many automobile and financial analysts foresee EVs making up no more than 10 percent of all new car sales over the next two decades.

Once again the "experts" are dead wrong. Most industry authorities predict EVs will make up just a few percent of new car sales a decade or more from now. Many, in fact, are fomenting fear and doubt with gloomy predictions of the electric grid buckling under the strain. Nothing could be further from the truth. Studies show that the current grid has the off-peak capacity to support as many as 180 million EVs. In addition, supply will grow with demand as energy sources increase and the infrastructure to support the delivery of that energy is deployed. One great thing about electricity is that there are so many ways to make it. Ultimately, in fact, far from collapsing the grid, electric vehicles will contribute to its growth, with the vehicles themselves feeding stored energy back into the system when it needs it most.

So why not buy an electric car? Why not adopt something that is better, faster, and cheaper? There's no reason not to, which is why Americans will buy EVs, and why demand will soar. With only a few modifications to the existing grid we can already replace 70 percent of the cars on American roads with electric vehicles. As for the remainder—well, hasn't American capitalism always responded to demand?

Both the Internet and the EV are shining examples of the efficiency of moving electrons rather than molecules. When a new technology emerges that is significantly better, faster, and cheaper—from eBay to EVs—everything else falls into place. America's entrepreneurial spirit has always managed to support market demands, no matter the perceived obstacles. Although Internet naysayers insisted there wouldn't be enough bandwidth to accommodate dramatic growth—that the Internet would essentially grind to a halt under the weight of so many users—the opposite occurred. Companies responded to the increased load by creating more and faster network connections. More and faster network connections in turn led to more and varied content and services, which led to more users, and so on, creating a cycle of growth. When connections increase, user experience improves. Google, for example, wouldn't be very useful if it were able to search just a few websites.

Similarly, as more public charging stations come online over the next couple of years, even those consumers who are generally slow to adopt new technologies will become increasingly comfortable with the idea of buying an EV. More electric vehicle purchases will lead to more charging stations, which will lead to more EV buyers, and so on. The key to the entire system is capitalism. While research grants and market incentives have been and will continue to be helpful in accelerating the growth of the EV market, the mass-market dynamics that follow on their heels—free enterprise in action—will dwarf all other mechanisms. As with the Internet before it, EV growth will be organic, and the market will respond accordingly.

One point of comparison between the Internet and the EV yet to be determined is America's leadership role. The U.S was front and center when it came to the development of the Internet, which began with the U.S. Department of Defense in the years following Sputnik, and was only embraced by entrepreneurs much later. Most early EV innovations, particularly in the realm of battery advances, have also come from the U.S. In fact, for the past 150 years, every major technological market advance has come from the United States—including the automobile, the aeronautics and space industries, the computer, and the Internet, to name just a few. Innovation has always been a core tenet of America and its economy, with the Constitution the foundation for U.S. patent law. The Founding Fathers understood the importance of innovation at the very outset of the Industrial Revolution; Thomas Jefferson, in fact, granted the first American patent. And it was no accident that the original U.S. Patent Office was one of the three original buildings conceived and designed for the National Mall, occupying prime real estate between the White House and Congressional office buildings.

Today, however, China and other nations are pouring vast resources into developing the EV industry for their own benefit. If the U.S. wants to retain its technological lead, as well as capture a substantial piece of the EV electriconomy, it needs to invest in research, manufacturing, and infrastructure. To that end, the federal and state governments should shift dollars from less strategic expenditures and initiate tax breaks and incentives to stimulate the EV economy, just as China and other governments are already doing. Now is the time to encourage American consumers to make their next vehicle electric.

The price of a new car is generally the biggest barrier to consumers buying one. Because battery costs have yet to benefit from the full effects of volume and innovation curves, the initial purchase price of an EV is currently higher than an equivalently featured gas-powered car. Although a $7,500 federal tax credit brings EVs close to parity, tax credits are complicated and owners still have to shell out a lot of money upfront and then wait to get the refund. Smart people who do the math, however, will quickly realize that the overall cost of ownership of an EV is considerably lower than its ICE equivalent. Not to mention that you get several days of your life back each year as a result of no longer needing to find gas stations or sit around waiting for oil changes, brake jobs, and smog inspections. (Though it's more than likely that EV owners will spend at least that much time each year telling others just how cool it is to drive an electric car.)

As stated earlier, the federal government currently offers early adopters a tax credit totaling as much as $7,500 for the purchase of a plug-in car. Several states offer additional inducements—up to $5,000 in Texas, Georgia, and California, which also offers carpool-lane access to plug-in cars, even when occupied only by a single driver. Plug-ins may also be eligible for local incentives. Los Angeles, for example, will subsidize charger installation up to $2,000 for the first 5,000 residential customers. Washington State Assembly Bill 1481, passed in 2009, contains provisions designed to speed the adoption of electric vehicles, including requiring that alternative energy sources fuel all publicly owned vehicles by 2015. The bill also encourages development of the infrastructure to support them, with model ordinances being streamlined to speed deployment. And Oregon residents can obtain online permits for Level II home charging station installation, with authorized contractors able to install the station and sign off on it, with no inspection required.

Many other states and municipalities, however, need to come on board to encourage faster adoption of EVs. Once electric vehicles become mainstream, with the infrastructure to support them firmly in place, subsidies will no longer be necessary. But at this stage of the game, they're essential. EV subsidies also level the playing field with oil companies, which currently receive subsidies totaling billions of dollars annually for exploration and drilling, as well as U.S. military support for the protection of global oil production and supply lines. Thomas Friedman of *The New York Times* has repeatedly used his column to recommend a $1 "Patriot Tax" be applied to gasoline. Given that U.S. gas prices—after adjusting for wages—are the lowest in the world outside of Saudi Arabia and Venezuela, Friedman's approach is probably the quickest and most effective way to achieve a more secure and financially robust nation.

Some have argued that a gas tax would stifle the economy. The facts, however, show otherwise, according to Gustavo Collantes, senior energy policy adviser with the Washington State Department of Commerce. Far from triggering an economic decline, Collantes argues that taxing gasoline will actually result in growth. For that to happen, however, the money needs to be recycled back into the economy through funding for electric vehicles and infrastructure, as well as tax credits and reductions. Collantes also perceives higher gas taxes as a way to level the playing field, since so many of the costs in gas are hidden.

Calling the Nissan LEAF a "game-changer," Collantes sees the vehicle an important step toward digging America out from underneath the tremendous debt burden it's building, a massive part of which comes from oil imports. "We have a saying in Spanish," said Collantes, who is originally from Argentina. *"Una espada sobre la cabeza"*—a sword hanging over our heads, ready to fall without warning. "We don't see the consequences of all this borrowing," he continued. "We don't see the impact of debt on future generations. But we need to see it, and we need to do something different now. One way is to reduce oil imports."

As we saw in the post-Sputnik era, national leadership is required to initiate policy and direction. But leadership at the local government and grassroots levels, the kind of leadership Mayor Annise Parker practices, is also necessary. While there is no stopping an idea and a technology whose time have come—widespread adoption of the electric car is inevitable—it would be even better if the leaders of this nation worked together to speed the process.

Just as certain policies stimulate nascent economies and enable change, others have the opposite effect, blocking growth and slowing adoption. With that in mind, policy needs for the emerging ecosystem include:

- Tax breaks for businesses of all sizes involved in advancing the EV ecosystem

- Government-funded research for energy storage

- Reduced barriers to entrepreneurship

- Acceleration and consolidation of technology standards

- Accelerated approval at the U.S. Patent Office regarding innovations related to energy and electric vehicles

- Streamlined building codes to remove barriers to the development of energy-generation infrastructure and EV public charging infrastructure

- Enactment of a federal mandate for the design and deployment of an interstate EV charging grid

The years between now and 2015 will see the launch of the next 50-year market—perhaps even 100-year market. The EV and its ancillary industries are going to be big, and now is the time to get up on the steamroller or risk becoming part of the road.

"The aeroplane will never fly."

—Lord Haldane,
British Secretary of State for War, 1907

CHAPTER 23

Electric Overdrive

"No flying machine will ever fly from New York to Paris." So stated Orville Wright, despite having successfully collaborated with his brother, Wilbur, to build and launch the world's first airplane in 1903. More often than not, the full impact of paradigm-shifting breakthroughs are missed, even by their inventors, and the Wright brothers hardly stand alone in their lack of long-term vision. Few anticipated the market's embrace of the PC, the Internet, or the cell phone either. Thus far, the same is true of electric vehicles, which are a lot closer to becoming an everyday reality than the public may realize. Over the next few years it will become increasingly evident to all Americans that the future of personal transportation will not only look a whole lot different, it will also be phenomenally better, faster, and cheaper.

What follows is a look at the three periods of EV development and adoption that we'll see over the next two decades.

2011-2020

One obstacle to mass consumer adoption of plug-in cars today is the limited availability of a public charging infrastructure, though as noted in the previous chapter, this is less of an obstacle than it seems. While there are as many as 250 million cars in America, those cars have only between 50 million and 70 million garages to call home each night. Some 40 percent of U.S. residences don't even have a carport, according to Coulomb Technologies, which notes that in San Francisco only one in six cars is parked in a garage. That leaves a lot of drivers who live in apartments and condominiums dependent on shared residential and street parking for their charging needs. Where will they charge their cars?

As electric vehicles become more common throughout the latter half of the decade, the market will grow to meet the demand. Level II charging will become increasingly widespread along city streets, in workplace parking lots and garages, and in commercial areas, allowing drivers to charge or even just top up their batteries while they work, eat, or shop. Quick-charge Level III charging, comparable to filling up at a gas station and capable of boosting a battery to 80 percent charged in 20 minutes or less, will also become more widely available across the nation's interstate highway system and in urban areas.

While there are more than 25,000 public EV charging stations already deployed, or scheduled to be deployed, within the next year, by 2016 there will be more public electric car charging stations in the U.S. than gas stations. As a result, by 2020, consumers will be able to drive all-electric vehicles from coast to coast, taking 20-minute fueling breaks every 300 to 400 miles—just as they do with their internal combustion engine cars today.

And while the electric motors powering EVs will see a 25 percent improvement in efficiency between now and 2020—going from 4 miles per kilowatt hour to 5 miles per kilowatt hour—Moore's Law's assures far more technology headroom when it comes to storage technology. In 1965, Gordon Moore, co-founder of Intel, observed that that the number of transistors per square inch on integrated circuits had doubled each year since the integrated circuit was invented, and predicted that the trend would continue for the foreseeable future. Moore's Law explains why computers and other electronics double in capacity and speed every couple of years, even as they fall dramatically in price.

PLUMMETING ELECTRICITY STORAGE COSTS

Reductions in Battery Price Over the Next 20 Years

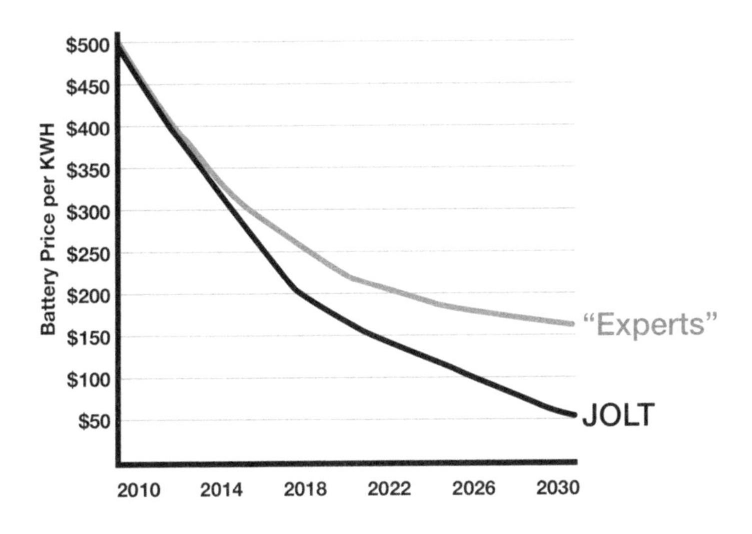

In terms of price and performance, batteries have historically improved at a rate of 8 to 10 percent per year. To put that in perspective: if a gas-powered car getting 20 MPG in 1990 were to improve at a rate of 10 percent per year, it would get 122 MPG in 2010. Given the recent increase in R&D spending for batteries and other electricity storage technologies, JOLT! *predicts that the rate of improvement will exceed 10 percent per year, bringing the cost per kilowatt hour of an EV battery down to just over $50 in 2030.*

While Moore's Law does not directly apply to all aspects of EV and energy storage technology, much of what does drive the cost and performance of electric vehicles is science and electronics. High rates of improvement as applied to electric vehicle technologies means that EV power storage units during the next decade will enjoy rapid improvement in terms of weight, volume, power, charging efficiency, cost, and lifetime use. And while it is possible that potential breakthroughs could result in batteries and ultracapacitors improving by 10 times or more by 2020, the known technologies being proven in labs today will take us to a factor of three—or three times today's capabilities. By 2020, a battery pack that today weighs 400 pounds, costs $12,000, and provides a range of 100 miles will cost less than $5,000 and travel 300 miles—and that is with no major breakthroughs.

If, however, companies such as A123 do indeed achieve major technology breakthroughs in the coming decade, we could well see a 300- to 400-pound battery capable of taking a car as much as 500 miles by 2020. That's about 10 hours of driving time, which is more than most human bladders or derrières can stand without a rest stop and quick charge.

Technological superiority aside, EVs will become increasingly popular as people see their neighbors and colleagues pull up in electric cars. After being forced to hear how fun the car is to drive, how great it is not to have to buy gas, how quiet the EV is, and how quickly it accelerates, their most likely response will be to go out and buy an electric vehicle for themselves.

And while economics and functionality are critical when it comes to marketplace success, nothing trumps "cool." For the next 10 years, EVs will be the epitome of mainstream cool—the newest must-have acquisition, their earlier image as a strange, new-tech purchase firmly in the past. By the end of the decade, however, plug-in cars will have run through that phase, and driving on gasoline will be seen as downright dumb, just as smoking is today. Even more, by 2020 drivers of gasoline-powered cars will be viewed as un-American. They'll be seen as anti-social, unwilling to do their part to help the country rid itself of its dependence on foreign oil and put an end to sending money to countries that fund terrorism.

2021-2030

Why would anyone, even a technology laggard, buy a gasoline-powered car after 2020? The better-faster-cheaper model will be so obvious at this point in the development of the electric car that only the most obstinate stick-in-the-mud would choose otherwise. By 2021, in fact, there will be no reason not to buy an EV.

IT'S 2020: WHY WOULD ANYONE BUY A GAS CAR?

2020: Features and Costs Comparison—ICE vs. EV

Feature	ICEV	EV
Purchase price	$25K	$25K
0-60	10 sec	8 sec
Top speed	110	110
Range	300	300
Safety	5 stars	5 stars
Weight	2200	2000
Refueling time	10 min	20 min
GHE	.18 kg CO2/km	.05 kg CO2/km
Maintenance/Year	$743	$126
Miles/Fuel dollar	5	70

Even with no major technological breakthroughs, by 2020 the many advantages of an EV over a gas-powered car will be undeniable, even to the most stubborn holdouts.

Throughout this next era, plug-in vehicles will become the new normal, gas-powered hybrids will fade away, and the paradigm shift of using electrons rather than molecules to power cars will be complete. Ongoing advances in energy storage technology (Moore's Law again) will see EVs capable of traveling as far as gas-powered cars on one charge, and all for less money. Drivers spending 20 or 30 cents per mile for emission-producing gasoline will be viewed as eccentrics and oddballs, just as early-EV adopters once were.

The 2020s represent the tipping point, when EVs will move from making up less than 10 percent of the vehicles on the road to mass adoption (90 percent of new vehicle sales) as natural attrition sees the vast majority of gas-fueled cars retired and replaced with plug-ins by the end of the decade. As a result, yet another law kicks in at this point: Metcalfe's Law posits that the utility of a network goes up exponentially with the number of users. As noted in the last chapter, Google isn't much good if it only has a few websites to search, just as phones are only useful in proportion to the number of other people that have them.

Although based more on theory than actual science, Metcalfe's Law is widely accepted by business planners and strategists, and has implications for the EV in terms of how the charging infrastructure will roll out. If one person in a neighborhood or town has an EV, there isn't much incentive for a business to install a public charging station. Once 10 people in the area have electric cars, however, the equation changes, and the infrastructure starts to expand. Ten new EV drivers in a different area leads to yet another charging station, and so on, and soon enough there are stations linking the different regions. A large number of charging stations throughout a region or along a heavily trafficked corridor has a psychological effect as well, resulting in increased confidence not just among EV owners, but also among drivers of gas-powered cars who are considering going electric. That confidence, boosted by ongoing advances in battery technology and growing familiarity with plug-in vehicles, will have a viral effect on EV sales.

Another big change during this period concerns energy storage. Today, the grid has no storage capacity. Any power flowing around the grid that doesn't have anywhere to go is pushed into the ground and lost. By the end of this era, however, that will no longer be the case, and the ability to store electricity will change everything, essentially turning our homes into micro-power plants, able to capture and use energy from alternative sources, just as Roger Brent does today. Fully charged EVs connected to the electrical grid will become extensions of it, supplying utilities with stored energy as needed during periods of high demand. Utilities will be able to avoid brownouts on hot summer days, better anticipate and manage power use, and limit the number of new power plants brought online. This vehicle-to-grid function will make cars useful even when idle, and could also provide EV owners with additional income or credit in the form of a utility payback.

One future scenario even envisions energy caching as the gas station of the future. A dumpster-sized battery pack parked in a garage, gas station, or parking lot will capture energy—solar, wind, or from the grid each night when demand is low—which will then be available to be dumped into the car in about 10 minutes. And since the pack will be stationary, size and weight won't be an issue, making it a perfect way to reuse EV batteries.

2031 and Beyond

If children of the previous era will view the few remaining gas-powered vehicles on the road quizzically, young people after 2030 will be incredulous to hear that we once wasted a nonrenewable fossil fuel on transportation. With all the useful applications for petroleum—aspirin and other pharmaceuticals, fertilizers, pesticides, super plastics, paintbrushes, balloons, tires, lunchboxes, computers, surfboards, and who knows what else by then—why in the world, they'll wonder, did we once burn it up just to move cars?

"God helps them that help themselves."

—Ben Franklin

CHAPTER 24

A Tale of Two Cars

In 2009, 356,800 Americans bought a new Toyota Camry, making it the top-selling car in the U.S. That same year, 290,100 other Americans bought Honda Accords, and 180,400 more bought Ford Fusions. These three models alone sell about 2,300 cars every day of the year, or about 100 every hour.

Each Camry, Accord, and Fusion buyer made an eminently practical choice: the sedans all start at around $20,000 and deliver 24 miles per gallon in combined city/highway driving. What could be more sensible?

An electric vehicle, of course.

But don't just take *JOLT!'s* word. Let's break down the numbers. Let's see just how rational the choice of a PHEV or EV will actually be in late 2010.

We're not going to compare our plug-in with a Chevy Suburban, though. Instead, let's compare two relatively equivalent cars:

- One is a "composite" traditional gas-powered Toyota Camry/Honda Accord/Ford Focus (we'll call it a Foyonda)

- The other is a "composite" of the PHEVs and EVs soon coming to market, including the GM-Volt, the Nissan LEAF, and the Ford Focus (we'll call it a Volteacus)

Next, let's make a few basic assumptions and see what owning these two cars might cost over five years:

	Foyonda	Volteacus	Comments
Powertrain	Internal combustion	Plug-in hybrid or all-battery electric	
Sticker price	$25,000	$35,000	Average sticker prices predicted for fully equipped cars
Federal tax credit	0	$7,500	See DOE information below*
Average additional incentives/tax breaks	0	$2,000	This can be as high as $5,000 in some states
Net price	$25,000	$25,500	EVs are close to parity after current incentives
Average fuel economy	8 miles/ dollar	50 miles/dollar	PHEV/EV blended average
Annual miles driven	15,000	15,000	National average for family sedans
Cost of fuel/year	$1,875	$300	$3.00/gallon $0.09/kilowatt hour
Cost of fuel/5 years	$9,375	$1500	$3.00/gallon $0.09/kilowatt hour
Maintenance/year	$743	$126	Tune-up, oil changes, repairs, etc.
Maintenance/5 years	$3,715	$630	
Subtotal cost for 5 years	$38,090	$27,630	$10,460 advantage—Volteacus

*For more information on U.S. Department of Energy tax credits for plug-in hybrid electric vehicles, go to http://www.energy.gov/taxbreaks.htm.

Although grossly oversimplified, the table above clearly shows the five-year cost advantages of the EV, in which the higher purchase price of the EV is recovered in incentives, fuel, and maintenance cost savings.

But it doesn't end there:

- Gas prices: Do you believe that the cost of gas will stay flat for the next five years? In the past five years the national average cost for a gallon of regular unleaded has gone from $2.10 to $2.86, a jump of 35 percent. The five years from 2000 to 2005 saw a rise from $1.50 to $2.10—that's a 40 percent hike. And will we ever forget July 2008, when we paid $4.12 a gallon, a hike of 95 percent over the 2005 price? But even if we use the past five-year jump of 35 percent as our guide, you're still likely to see the total average cost of the gas you'll buy over the next five years climb by about 21 percent. Now take a look at the chart below to see the cost savings our Volteacus provides simply because it doesn't need much—if any—gas.

	Foyonda	Volteacus	Comments
Powertrain	Internal combustion	Plug-in hybrid or battery electric	
Sticker price	$25,000	$35,000	Fully equipped
5-year subtotal	$38,090	$27,630	$10,460 advantage—Volteacus
Additional fuel cost for 5 years, assuming 21 percent inflation	$1,969	$137	$1,832 advantage—Volteacus
Subtotal cost for 5 years	$40,059	$27,767	$12,292 advantage—Volteacus

- Resale value: A fully loaded, good-condition five-year-old Camry with 75,000 miles on it has a trade-in value of $8,500 today, according to Kelley Blue Book. (That price reflects a relatively strong Toyota value retention of 34 percent.) By 2015, there will be hundreds of thousands—if not millions—of PHEVs and EVs on U.S. roads. The charging infrastructure will be ubiquitous, and range anxiety will seem quaint. The reliability and usability of the cars with plugs will be proven. And oil and gas prices will almost certainly continue to climb. So which of our two 2010 choices is likely to command the higher resale value?

- Maintenance costs: The carmakers are all predicting that the costs of maintaining their PHEVs and EVs will be a lot lower than traditional gas-powered cars. Even plug-in hybrids, with their much smaller and less-used engines will cost less to maintain and service than typical internal combustion cars. And don't forget that brakes on plug-in cars last far longer than brakes on traditional cars. Electric vehicle brakes are actually induction motors, turning the car's kinetic energy into electricity rather than brake dust and heat.

- Hassle factor: Gas cars need filling up, and how long does that take? Maybe 10 minutes? That's not so much. But if you fill up every 300 miles, that works out to 50 times a year—500 minutes, or 8.33 hours, per year. That's about 42 hours of your life over 5 years. Not to mention the time spent having the internal combustion car serviced. How much do you enjoy that chore? (Conservatively, we can estimate the gas car goes to the dealer twice a year, while the plug-in goes once. Maybe.)

- Bragging rights (or simply quiet pride): Who doesn't want to import/drill for/burn less petroleum? Even our big oil companies are advertising their commitment to alternative and sustainable sources of power. Visualize showing your new electric car to your good neighbors. Imagine the thoughtful conversation you'll have about what you've learned concerning the future of energy and the electrification of transportation. Picture their reactions once they experience the remarkable performance of your twenty-first century car.

- Fun: The torque of the electric motor is max at 0 RPM, which translates to rapid acceleration—and fun. Of course, this concept is a leap of faith until you drive your very first car with an electric motor attached to the wheels. Once you have, however, you will thank the laws of physics that your EV gives you both better mileage and better performance. Doesn't seem fair, but there it is.

Of course, it's challenging to create an accurate cost comparison between cars we're familiar with and the new alternatives—what we know and what we don't yet know. Any comparison requires assumptions about car prices and fuel costs, and those assumptions can tilt the outcome. Nonetheless, current trends clearly favor cars with plugs. And the giant carmakers know it, too, which is why they have irreversibly committed themselves, spending billions to bring the new cars to market.

The point of this chapter was to use our Foyonda and Volteacus to make the basic case for buying a plug-in car. The bottom line? The Volteacus saves time and hassle, provides an awesome (even historic) conversation starter, and is a lot of fun to drive.

And if you care nothing about any of those things, here's the kicker: it will save you money.

"Vision without execution is hallucination."

—Albert Einstein

CHAPTER 25

NOW WHAT?

Electric vehicles are our future no matter who builds them. The undeniable reality of a better, faster, and cheaper technology means we will all be driving EVs within the next 20 years. But to ensure plug-in cars are part of America's economic future as well, it's time to take a page from our Sputnik past, when Americans pulled together to defeat a seemingly overwhelming threat to our national security—and in the process achieved spectacular technological advances and economic rewards.

By now it should be clear that electric vehicles are about far more than saving the environment. Aside from the fact that EVs are a lot of fun to drive, they're also about energy independence. They're about keeping America strong. They're about saving—and creating—American jobs. As Ray Lane of Kleiner Perkins points out in Chapter 12, if you completely take the environmental issue out of the equation, you still end up with a phenomenal business opportunity. In a world that will grow from 6.8 billion people to 9.2 billion by 2050, demand for renewable energy is going to skyrocket. Electric vehicles and other forms of alternative energy technologies are the next great global industries.

"America still has the best innovation culture in the world," Thomas Friedman wrote in his *New York Times* column in March 2010. But innovation isn't enough. We need better policies to nurture that innovation, he said. We need better infrastructure to enable it.

So how do we nurture that innovation? How do we enable it?

We need a concrete and far-reaching national energy policy and plan, one that will take us down the road to energy independence and the electrification of our passenger cars and trucks. And the best way for us to get there is for our president to take a hard stand. Just as President Kennedy did nearly 50 years ago, when he challenged the American people to send a man to the Moon, President Obama must face the American people and issue a challenge of his own: an electrified, energy-independent America by the end of this decade. If this great nation could make it to the Moon and back in less than 10 years, we can certainly achieve energy independence within the same timeframe.

The president needs to set a direction and tone for America, telling the nation that now is the time for us to end our oil addiction by adopting cars that run on electricity. Now is the time to end the insane practice of sending nearly $500 billion overseas each year to pay for foreign oil—money that too often ends up funding the wrong side of the war on terrorism. Instead, we need to use our hard-earned money to build America—American schools, American infrastructure, American research, and American economic might.

Now is also the time to initiate energy and business policies that will stimulate the electric vehicle and other clean tech industries. It's time to throw our money and muscle into research and the implementation of alternative energies. It's time to offer incentives and tax breaks to businesses—just as China does—to build factories here, in America, rather than overseas. And it's time to inspire our teachers and children, particularly in the fields of science and engineering, just as we did in the years following Sputnik.

Now is the time to build the twenty-first-century smart grid, capable of securely supporting and advancing the electric vehicles that will drive our future. Now is the time to implement U.S. Energy Secretary Steven Chu's modern-day Manhattan Project. Instead of using America's best minds to build weapons—or to trade derivatives on Wall Street—we must bring those minds together to create and develop the next generation of storage devices to power our next-generation cars, homes, and industries.

EVs are the future, and the future is here. Now is the time to make it ours.

In that spirit of cooperation and nation-building, here then is the "JOLT! Program for America," which calls for 100 million plug-in cars on American roads by the end of the decade. To achieve that goal, *JOLT!* recommends the following objectives and milestones:

- An additional 25 percent tax on all oil imported from OPEC nations, with the proceeds used to fund the following incentives and grants for electric vehicle and EV infrastructure development

- Since the federal government buys more cars than any national entity or organization, 50 percent of its light-duty fleet should be electric by 2015, and 90 percent electric by 2020; all state government light-duty fleets should achieve that same target as well

- 50 percent corporate tax credit for all electric vehicle and associated infrastructure expenditures through 2015

- A 100 percent federal tax exemption for all corporate revenue produced from EV charging infrastructure, battery production, and battery recycling

- A doubling of all tax credits for corporate R&D expenditures for EV and related electric power generation and distribution through 2015

- Venture capital tax credits for EV and related technology investments

- A lithium exploration and import tax credit, as well as a lithium export tax, since the goal is to keep all lithium stocks in the country for recycling purposes

- $10 billion in national laboratory and university grants to fund EV and all related electric power generation and distribution research through 2015, including a NASA-like program dedicated to grid infrastructure development

- $1 billion in federal money to fund college scholarships for top American students pursuing engineering and science degrees related to EV technologies

- A $10,000 annual tax credit for the creation of qualified jobs in the EV and associated industries through 2015

In addition, *JOLT!* recommends the following consumer incentives to stimulate EV adoption through 2015:

- An immediate government rebate of up to $10,000 on all electric vehicle purchases ($500 per kilowatt hour of battery capacity), as well as an additional $2,500 "cash for clunkers" trade up from a gas-powered car to a plug-in

- No state sales tax for plug-in vehicles

- Carpool lane access for all plug-ins, even those occupied by single drivers

- Free access to battery-powered public transit

- Up to $2,000 in tax breaks for Level II home-charger installation

And because hard work deserves recognition and acclaim, *JOLT!* also offers up the "JOLT! AMERICA Awards"—initially funded from the net proceeds of this book—which will be given in recognition of outstanding efforts in creating, enabling, and promoting an EV world, from innovation and research to education and government initiative. Categories will include:

- Educators from the intermediate, high school, and college levels
- Universities and research centers
- Individual innovators
- Small companies
- Large corporations
- Venture capital firms
- Congressional leaders
- State leaders
- Local leaders
- JOLT! AMERICA Person of the Year

To prove what a transcendent issue EVs are, the first annual event should be televised and co-hosted by the likes of Jon Stewart, Keith Olbermann, Rush Limbaugh, and Glenn Beck. If such outspoken and politically opposite media personalities can come together to honor some of our country's best and brightest—those committed to moving our nation forward—it will be patently clear that nothing and no one can stop us.

Visit www.joltthebook.org for details.

These are some of the things we can and should do to take charge as a nation, and what JOLT! intends to do as an organization. But what about you?

What can you do to help build an electric-based transportation system that is significantly better, faster, and cheaper?

What can you do to help our great nation end its dependence on oil and all the problems that come with that dependence?

What can you do to help clear the air and ensure that American waters never again suffer a horrific oil spill that results in devastating economic and environmental effects?

What can you do to help assure that the benefits of the electriconomy remain at home, rather than being exported overseas, the way so many oil profits are today?

What can you do to help keep America safe, strong, and prosperous?

What can you do to take charge?

It's actually pretty simple—and potentially a lot of fun. You can start by buying an EV (that's the really fun part), or getting your name on a waiting list to signal your interest and support. But even if a new car isn't in your budget, there is plenty you can do to help build an EV world and develop an American-led electriconomy:

- Vote and campaign for politicians with EV plans and a commitment to see them initiated. Go to www.joltthebook.org to see how JOLT! rates many of America's leading public officials.

- Pursue a career in EVs or other forms of alternative energies

- If you have money to invest, invest in EVs and other forms of alternative energies

- Petition your utility company to invest in American-made renewable energy sources

- Evangelize. Tell everyone you know about electric vehicles, and encourage them to purchase EVs and vote for politicians who support them.

- Go to www.joltthebook.org and download a PDF copy of the JOLT! PowerPoint presentation on EVs, which you can then present yourself to groups of friends, co-workers, schools, community centers, and local businesses

- Send ideas on charging America to myproposal@joltthebook.org

And while these suggestions might sound simplistic, they're not. Each is a step toward oil independence. Each is a step toward building a stronger, more economically vibrant United States of America. Each is a way to make a difference.

America, it's time to take charge!

"You can tell whether a man is clever by his answers. You can tell whether a man is wise by his questions."

—Naguib Mahfouz, Nobel Prize-winning author

ADDENDUM

Roadblocks and Speed Bumps

(Frequently Asked Questions)

Although consumer surveys show strong interest in electric vehicles—the Nissan LEAF sold out its first-year production line over the Internet without the car ever seeing the inside of a showroom—there remains a general lack of knowledge about EVs and how they fit America's car culture. Below are answers to common questions and concerns.

Doesn't the limited range of an EV pose a problem?

Even though EVs travel approximately 100 miles on a charge and the average American's commute is less than 30 miles, drivers still worry about being caught on the road with a dead battery. Given that more than 90 percent of all trips are less than 30 miles, however, range is less of a problem than many think. Additionally, all EVs come with a "travel cord" that enables the driver to plug into any standard 110-volt outlet. Still, today's EVs aren't appropriate for longer-haul trips, which is why plug-in hybrids, which go between 10 and 40 miles on an electric charge before the gas engine kicks in, are likely to make up a significant number of sales for the next few years. Many people will buy pure EVs as second vehicles. Concerns around range will dissipate with the introduction of Level II and Level III chargers throughout communities and along highways. In addition, advances in battery technology should make the EV's range equivalent to that of gas-powered vehicles within a decade. Efforts are also underway to establish mobile charging units, which stranded motorists could call for a "refill."

Don't EVs take a long time to charge?

Not really. A Level II charger can fill the battery of a 100-mile range car in under 4 hours and a Level III charger can get you 80 more miles in under 20 minutes. And both those stats assume the battery is completely dead, which will almost never be the case. Since U. S. Department of Transportation statistics show that the average car is parked at home 75 percent of the time, including overnight hours, charging the battery (even at Level I speeds, which requires only a 120-volt outlet) isn't much of an issue, though many homeowners will choose to upgrade to Level II (see entry below). A study of more than 1,000 EVs in Tokyo showed that 90 percent of the time the batteries were more than half full when plugged in for a new charging session. Most EV drivers won't run their batteries to "E" before plugging in and are likely to "top off" their vehicles whenever they pull into their garages or pull up to a charging station. In addition, as the public charging infrastructure is developed and expanded, Level II charging will become increasingly available at work and in commercial areas.

Will I need to install charging infrastructure at home?

If you have a power outlet in your garage or parking shelter, your home charging infrastructure is already in place. Outlets providing 120-volt outlets enable Level I charging, which requires about eight hours for a complete charge; 240-volt outlets allow Level II charging, which takes just a few hours, depending on the battery's initial state of charge. As electric vehicles become more common, public charging stations offering both

Level II and Level III chargers (Level III chargers provide a quick charge, comparable to filling up at a gas station) will grow to meet the demand. Government grants and incentives will also play a role in spurring the deployment of public charging infrastructure.

Doesn't it cost a lot to install a charging system at home?

Level I charging requires nothing in the way of installation. Plug the car into a standard 120-volt outlet and you're done. Homeowners, however, may choose to upgrade to Level II charging, which requires 240 volts, the same voltage as a standard clothes dryer. Upgrading the system, which requires an electrician, ranges in cost from a few hundred dollars to $2,000. In many cases, federal and state incentives and tax breaks cover most or all of this cost.

How will I charge my car if I don't have a garage or if I need to top up my battery while away from home?

There are nearly 250 million vehicles in the U.S., and more than half of these cars are housed in the nation's 50 to 70 million private garages (many with more than one spot). In addition, most people who buy new cars have garages, so charging is less of a problem than most people think. Those who do park their cars on the street will need to rely on public charging stations, which will ultimately blanket city streets and parking lots. Auto companies are working with municipalities to find ways to finance Level II and Level III public charging stations. The networked intelligence of the charging stations will allow in-car telematics (telecommunication devices that send, receive, and store information) and access through smartphone applications, enabling drivers to find available stations and receive text messages when charging is complete.

Don't car batteries pose a safety risk?

Years of research and testing have gone into ensuring that the new lithium-ion batteries for cars (whose components are 95 percent recyclable) can tolerate temperature extremes and are well sealed and protected in the event of an accident. Since emergency workers are already trained in dealing with existing car batteries when responding to roadway accidents, a larger battery won't pose a problem. Vehicles loaded with gasoline—a highly explosive liquid—colliding at high speeds represent a much bigger danger.

Don't electric car batteries wear out and then require an expensive replacement?

The lithium-ion batteries in the latest electric vehicles are expected to last the lifespan of an average car in America—at least 10 years before their energy capacity falls below 80 percent, when they might need to be replaced. (They may, however, last longer than 10 years; many of the original NiMH batteries in RAV4 EVs are still going strong nearly 15 years later and, thus far, lithium-ion batteries appear to have a longer lifespan.) A battery no longer able to sufficiently power a car for the desired range is fully capable of being used elsewhere in what are known as "stationary applications," such as wind or solar power storage. As a result, the cost of a replacement battery will likely be partially offset by trading the old battery in for a new one.

Won't old EV batteries clog our landfills and cause toxic pollution?

No. Unlike gasoline, which is completely consumed in the process of extracting energy, lithium-ion batteries are more than 95 percent recyclable. A battery no longer able to maintain enough charge after 10 or 12 years to power the motor the desired number of miles can be traded in or sold directly to a battery broker, who can then sell it into a second life as an energy storage unit for stationary objects such as houses or windmills, which don't require automobile-grade capacity. And when a battery is ultimately discarded after another 10 or even 20 years, lithium and all the other components of the battery can be extracted and reused. Imagine if we only had to import enough oil to fill our gas tanks just once over the next 10 years! In addition, lithium-ion batteries are far less toxic than older lead acid batteries; nothing leaches out even when taken apart.

Will consumers really like EVs?

Consumers will love EVs and plug-in hybrids. Given the positive reception the Toyota Prius has received over the past few years (not to mention the fact that Nissan sold 19,000 LEAFs over the Internet well in advance of the car's late-2010 release), it's extremely likely that the many plug-in cars coming out over the next couple of years will be greeted enthusiastically as the public grows more familiar with them. Americans love cars, they love technology, and they love the next new thing. And the fact that celebrities such as Tom Hanks, George Clooney, and Leonardo DiCaprio are known to be avid fans of electric vehicles raises their cool factor even higher among segments of the public.

What effect will uncertain market conditions have on the EV market?

Electric vehicles are coming no matter what the state of the economy. A major crash in the next couple of years might slow the adoption of EVs, but it certainly won't stop it. GM, Ford, Mercedes Benz, Toyota, Nissan, and other carmakers have spent many years on electric vehicle R&D and have too much invested to stop. As Bob Lutz of GM has said, carmakers can and will make money on whatever vehicles the public wants to buy. A123 Systems, which has already banked hundreds of millions of dollars to fund lithium-ion battery research, is equally invested and isn't about to stop innovating either. And there are scores of other companies investing in EVs at an ever-accelerating rate. This trend is irreversible.

Won't electricity costs go up with increased demand?

Not at all. Unlike oil, electricity prices have remained stable and relatively inexpensive for the past 25 years, even in the face of greater demand. An increased number of renewable forms of energy coming online—solar, hydroelectric, wind, nuclear, etc.—will only contribute to price stability. In addition, electricity demand will rise among EV owners but will drop elsewhere as households gradually switch over to more energy-efficient appliances and lighting. Finally, since the technology exists to monitor what is being charged and when, utility companies can implement different pricing structures as a way of encouraging specific behaviors. A homeowner charging a car after midnight, for example, won't pay the same rate as someone heating a hot tub in the middle of the afternoon. Finally, it's possible that the increased demand for electricity driven by EVs will actually cause electric rates to fall thanks to economies of scale. Higher-volume production without the need for additional capital infrastructure or human resources most often means lower prices.

Don't EVs cost more than a gas-powered car?

EVs actually cost less to own than do gas-powered-cars. Spread over the lifetime of the vehicle, electric vehicles are far less expensive to operate than conventional cars—2 cents a mile for electricity versus 10 to 25 cents a mile for gas. Thanks to a regenerative braking system and a low-maintenance motor under the hood (and little else) they also cost far less to maintain. The additional 20 percent upfront cost associated with the EV is due to its battery, an expense that for early adopters can be offset by a federal tax credit for plug-in cars, which amounts to as much as $7,500. Some states also offer an additional incentive for initial sales ($5,000 in

California, Texas, and Georgia), and plug-ins may also be eligible for local utility incentives. (Los Angeles, for example, will subsidize charger installation up to $2,000 for the first 5,000 residential customers.) Technology advances and increased production of standardized units should drive the cost of new batteries down by half, or even more, over the next decade.

Won't the electrical grid collapse under the increased load?

Not even close. The U.S. electric grid operates at full capacity 24 hours a day. If most vehicles are charged overnight, when capacity is high and usage is low, the overall impact on the grid will be minimal. According to the U.S. Department of Energy, enough excess generating capacity exists at night to charge 180 million EVs without adding any new capacity in the form of natural gas, coal, or nuclear power plants. Eventually, EVs hooked up to the grid can supply extra power during times of peak demand, helping to prevent summer brownouts. In any case, modeling shows that the average daily charging requirement for an all-electric EV is expected to be between 10 and 12 kilowatt hours; a plug-in hybrid is expected to draw 7 or 8 kilowatt hours. This is because EV drivers tend to "top off" their vehicles whenever they pull into their garage or anywhere else they have access to electric fuel. EV drivers don't usually run their batteries to "E" before plugging in, though their behavior may change as both EV range and the number of public charging facilities increase, and range anxiety dissipates. Even with a 110-volt garage outlet, a plug-in car will be fully recharged during a night of parking and charging.

Don't EVs just transfer emissions from the tailpipe to the smokestack?

No. Even with 52 percent of all U.S. electricity generated from coal, an EV is far cleaner to operate than a gas-powered car. In addition, unlike conventional cars, plug-in vehicles will increasingly be run on renewable sources of energy such as solar, wind, and nuclear.

Conventional cars perform better than EVs, right?

No way. Just wait until you drive an EV! Electric vehicles are quiet and offer a high-performance driving experience. They are fun to drive, delivering greater power, torque, and acceleration than comparable gas-powered cars.

Is it true that EVs don't go as far in cold weather?

It's true that running accessories like the heater or air conditioner will reduce the mileage range of an EV. Because an electric motor is so efficient, it doesn't have a lot of excess heat to warm the interior of the car, so the battery is used to heat (and cool) the vehicle. The upside is that EVs are smart enough to pre-heat or pre-cool their interiors while still plugged into their electric chargers. Not only is no energy wasted to get the car to the desired temperature, but the driver also gets the benefit of comfort from the outset…Just another example of why the consumer experience with an EV is so much better.

Doesn't the EV's driving range shrink if I drive faster?

Just like a gas-powered car, EV mileage varies depending on driving style. Quick acceleration and higher-speed driving will reduce the expected range of an EV. However, the instrument panel gives the driver continuous feedback on how far the car can travel on the energy remaining in the battery. And don't forget that unlike in a gas-powered car, the EV braking system recaptures much of the lost energy and puts it back into the battery. In addition, no energy is used to turn the electric motor when the car is idling—another advantage over a conventional vehicle.

What about existing building codes?

The reality is that a Level II charger is a lot like a standard 220-volt dryer plug. As cities and states recognize the inevitability of EVs and realize they need to make it easier for motorists to go electric, changes to building codes are already taking place. In Washington State, for example, Assembly Bill 1481, passed in 2009, contains numerous provisions designed to speed the adoption of electric vehicles and the development of the infrastructure to support them. As more consumers demand and receive access to charging stations, residential and commercial building codes throughout the nation will be amended.

With fewer people buying gas, won't there be less money for road repairs?

On average, EVs are lighter than gas-powered cars and will cause less road wear. However, since gasoline taxes currently support road building and maintenance, the revenue will need to be made up through increasing the excise tax on gasoline and/or shifted to other sources. Options include taxing for miles driven, taxing electricity, and collecting fees for public parking spaces that are used for charging.

Won't lithium suppliers just become the new OPEC?

No. Not only does the U.S. have 24 percent of all known deposits of lithium, many of the other countries that supply it are friendly to U.S. national interests. Beyond ample supplies and stable prices, lithium is highly recyclable. Early investment in lithium-ion battery recycling will limit the need for raw lithium and reduce the need to import future supplies from other nations. Once we have a strategic hold on 1 billion pounds of lithium (4 to 5 pounds per car), America will need no new lithium for its fleet of 250 million electric vehicles. Even if no recycling were done, no new sites were found, and all mining stopped tomorrow, existing lithium stores would be sufficient to supply projected EV production for the next 75 years.

What about power outages?

America has one of the best power grids on the planet. Despite the occasional outage due to a storm, high demand, or human error, the grid very rarely goes down. When it does, it's generally only for a short period of time and in a confined area. Even if power is out in one neighborhood, it is likely that most drivers will have enough range to drive to a public charging station in a neighboring region (which they also have to do to fill up with gas, since pumps shut down when the power is out). Down the road, as homes increasingly become micro-power plants, relying on stored power and producing their own from solar panels and other renewable sources of energy, outages will become less disruptive, ultimately resulting in a stronger, more resilient grid.

Why should the government subsidize EVs?

We need to level the playing field. Don't forget that taxpayers currently subsidize oil and gas, with 15 percent of our military budget allocated to the protection of global oil production and supply lines. In addition, oil and gas companies receive generous government-funded credits. That said, tax incentives for electric vehicles and the development of their associated ecosystems will only be necessary in the early days of the EV revolution. Once the cars find acceptance, battery costs drop, and the charging infrastructure is up and running, subsidies can and should be eliminated.

Why can't we just drill for our own oil supplies?

There is nowhere near enough. Estimates of possible U.S. oil discoveries offshore and in the Arctic National Wildlife Refuge range from 0.6 billion to 13 billion barrels. These final possible sites for major U.S. oil discoveries, even if successfully tapped, would be insufficient to meet the country's demand, which is currently 20 million barrels a day, and would provide between just one month and two years worth of supplies. Even oilmen like T. Boone Pickens and George W. Bush have gone on record as saying there is not enough oil under American soil and oceans to come close to what it takes to run our economy. That fact, plus issues of national security, job loss, and the transfer of U.S. wealth—all a result of purchasing oil from other countries, many of which are hostile to our interests—demonstrate the good sense of combining what oil we do have with American-made electricity, leaving us dependent on no one.

WEBSITES FOR MORE INFORMATION

General information

JOLT! The Impending Dominance of the Electric Car and Why America Must Take Charge: www.joltthebook.org

Plug In America:
http://www.pluginamerica.org/

EV World:
http://www.evworld.com/index.cfm

Electric Auto Association:
http://eaaev.org/

All Cars Electric:
http://www.allcarselectric.com/

Electric Vehicle EV:
http://electricvehicleev.com/evnews.htm

U.S. Department of Energy:
www.energy.gov/

EV Finder (used EVs):
http://www.evfinder.com/about.htm

EV Charger News (EV supply equipment):
http://www.evchargernews.com/

Electric Vehicles of America (EV supply equipment):
http://www.ev-america.com/

Guide to EV conversion companies:
http: http://bit.ly/a2B7qF

Electric Cars Are For Girls (guide to EVs and conversions):
www.electric-cars-are-for-girls.com

Neighborhood Electric Vehicles:
http://autos.groups.yahoo.com/group/NEVs/

Center for Automotive Research:
http://www.cargroup.org/publications.html

The California Car Initiative:
www.CalCars.org

National Electric Drag Racing Association:
www.nedra.com

EVs and Plug-in Hybrids

Nissan LEAF:
www.NissanUSA.com

GM-Volt:
http://gm-volt.com/

Tesla Motors:
http://www.teslamotors.com

Ford Focus:
http://www.wired.com/autopia/2010/03/jay-leno-1909-baker-electric/

Fisker Karma:
http://karma.fiskerautomotive.com/

Coda Automotive:
http://www.codaautomotive.com

BYD Auto:
http://www.byd.com/

Energy Security

Set America Free Coalition:
www.setamericafree.org

Simmons & Company International:
http://www.simmonsco-intl.com/

Institute for the Analysis of Global Security:
http://www.iags.org/

Casualties of the Iraq and Afghanistan Wars

The Washington Post Faces of War:
http://projects.washingtonpost.com/fallen/

Operation Iraqi Freedom/Operation Enduring Freedom:
http://www.icasualties.org/

Climate Change

Intergovernmental Panel on Climate Change:
http://www.ipcc.ch/

Southern Alliance for Clean Energy:
http://www.cleanenergy.org/

U.S. Environmental Protection Agency:
www.epa.gov/

Union of Concerned Scientists:
http://www.ucsusa.org/

Natural Resources Defense Council (NRDC):
http://www.nrdc.org/

Sierra Club:
http://www.sierraclub.org/

GLOSSARY OF TERMS

Battery electric vehicle (BEV)—Also known as an EV, a vehicle propelled by an electric motor that draws energy from an on-board battery. BEVs produce no tailpipe emissions.

Capacitor—An alternative to batteries, a capacitor stores energy in an electric field. Because it produces no chemical reaction, it delivers energy quickly and forcefully, and can be charged in a matter of minutes, or even seconds. In addition, a capacitor can be charged thousands of times, giving it a much longer lifecycle than a battery. It also tolerates more extreme conditions, including heat, cold, and vibration, and unlike a battery, suffers no harmful effects from being completely discharged. Its capacity for storing energy, however, is limited, meaning it holds only a fraction of the power of a lead-acid or lithium-ion battery. While not ideal for powering an EV's drivetrain, capacitors can be utilized in components such as regenerative braking, air conditioning, and electric power steering, which require high-peak currents.

Drivetrain—Also known as the powertrain, the set of components that generate and transmit power to the wheels

Electric motor—An EV motor transforms electrical energy stored in the battery into mechanical energy

Electric vehicle (EV)—A vehicle propelled by an electric motor that draws energy from an on-board battery. EVs produce no tailpipe emissions

Grid-enabled vehicle—An electric or hybrid-electric vehicle that plugs into the electric grid to recharge an on-board battery

Hybrid-electric vehicle (HEV)—An HEV has an internal combustion engine and requires gas. Additional energy is stored in a battery, which powers an electric motor; that motor converts electric energy into mechanical energy. The two technologies work together to propel the car, resulting in reduced fuel consumption and tailpipe emissions.

Internal combustion engine (ICE)—A conventional vehicle that stores gasoline or diesel in a fuel tank. That fuel is combusted in the engine, which delivers mechanical energy to power the car.

Kilowatt (kW)—A measurement of electrical power equivalent to 1,000 watts, 1,000 joules per second, and about 1.34 horsepower. Signifies the power an electrical circuit can deliver to a battery.

Kilowatt hour (kWh)—A measurement of total electrical power used over time; 1,000 watts of power delivered for 1 hour

Light-duty vehicle—An automobile or light truck. Light-duty vehicles include cars, minivans, sport utility vehicles (SUVs), and pick-up trucks.

Nanocapacitor—A device offering dramatically increased energy storage density as well as high power. Its electrodes operate in the same way as those found in conventional capacitors (see above), but rather than being flat, they are tubular, giving nanocapacitors a larger surface area. Less chemically reactive than ultracapacitors, they are able to operate at a higher voltage.

Peak demand—The demand on the electrical grid during high-use times of the day or night

Plug-in hybrid vehicle (PHEV)—A PHEV is essentially a regular hybrid with a bigger battery and a smaller combustion engine, and a plug that works in a standard 120-volt or 240-volt outlet. PHEV batteries can power the vehicle in all-electric mode at modest speeds for 7 to 60 miles, depending on the size of the battery pack and driving conditions, before requiring assistance from the internal combustion engine. Once the battery pack is depleted, the gas engine is engaged to drive a generator to power the electric motor and extend the range of the vehicle. PHEVs get up to 100 miles per gallon. In a parallel PHEV such as the plug-in Prius, the wheels can be driven either by the gas engine, the electric motor, or both. In a serial PHEV such as the GM-Volt, the wheels are driven only by the electric motor.

Powertrain—See *drivetrain*

Regenerative braking—Release the accelerator and the electric motor slows the drivetrain, causing a reduction in speed. Energy captured from the slowing wheels in turn powers the motor, which acts as a generator, converting energy that is normally lost during coasting and braking into electricity, which the battery stores until needed by the electric motor.

Ultracapacitor—An enhanced capacitor (see above), with increased storage ability. Nonetheless, ultracapacitors store around 25 times less energy than a similarly sized lithium-ion battery.

Vehicle-to-grid platform (V2G)—Energy stored in electric vehicles that can be used to support the electrical grid during periods of high, or peak, demand

Zero-emissions vehicle (ZEV)—A vehicle that produces no tailpipe emissions

ACKNOWLEDGMENTS

In my experience, nothing good is accomplished without the help of others. *JOLT!* is no exception—far from it.

First and foremost, thank you to Joseph DiNucci, my good friend and close colleague of more than 25 years, who introduced me to the state-of-the-art in electric vehicle technology several years ago. Typical of Joe's style of total immersion and inclusion, he brought me up to speed on today's next-generation cars, and then remained with me every step of the way throughout the writing of *JOLT!*

I am also grateful to Joe for introducing me to the people of Azipan: Kate Benediktsson, Lauren Cuthbert, and Atiya Davidson. They have worked tirelessly for the past six months to make this book great and keep it on schedule.

Thank you to Jonathan Broadus, who did a marvelous job on the cover art and graphics, and to Adam Witty of Advantage Media Group and Karen Ammond of KBC Media for their help and guidance.

A very special thank you to my brother, David Billmaier, who two years ago helped me create the original PowerPoint presentation, which became the basis for *JOLT!* He has been a trusted advisor throughout this entire process.

Finally, I want to express my heartfelt gratitude to the American leaders and entrepreneurs interviewed for *JOLT!* They are dedicating their lives to making this country a better and stronger nation.

Advantage Media Group is proud to be a part of the Tree Neutral™ program. Tree Neutral offsets the number of trees consumed in the production and printing of this book by taking proactive steps such as planting trees in direct proportion to the number of trees used to print books. To learn more about Tree Neutral, please visit **www.treeneutral.com.** To learn more about Advantage Media Group's commitment to being a responsible steward of the environment, please visit **www.advantagefamily.com/green**

JOLT! is available in bulk quantities at special discounts for corporate, institutional, and educational purposes. To learn more about the special programs Advantage Media Group offers, please visit **www.KaizenUniversity.com** or call 1.866.775.1696.

Advantage Media Group is a leading publisher of business, motivation, and self-help authors. Do you have a manuscript or book idea that you would like to have considered for publication? Please visit **www.amgbook.com**

How can you use this book?

MOTIVATE

EDUCATE

THANK

INSPIRE

PROMOTE

CONNECT

Why have a custom version of *JOLT!*?

- Build personal bonds with customers, prospects, employees, donors, and key constituencies
- Develop a long-lasting reminder of your event, milestone, or celebration
- Provide a keepsake that inspires change in behavior and change in lives
- Deliver the ultimate "thank you" gift that remains on coffee tables and bookshelves
- Generate the "wow" factor

Books are thoughtful gifts that provide a genuine sentiment that other promotional items cannot express. They promote employee discussions and interaction, reinforce an event's meaning or location, and they make a lasting impression. Use your book to say "Thank You" and show people that you care.

Jolt! is available in bulk quantities and in customized versions at special discounts for corporate, institutional, and educational purposes. To learn more please contact our Special Sales team at:

843.300.4980 • sales@advantageww.com • wwwAdvantageSpecialSales.com